S0-ACM-931

QUEUEING THEORY:
A Problem Solving Approach

Leonard Gorney
King's College, Wilkes-Barre

PBI
a petrocelli
book
new york / princeton

T
57
. 9
. G67
1981

Fordham University
LIBRARY
AT
LINCOLN CENTER
New York, N. Y.

Copyright © Petrocelli Books, Inc.

All rights reserved.
Printed in United States of America
1 2 3 4 5 6 7 8 9 10
Typesetting by Backes Graphics

WITHDRAWN
FROM
COLLECTION

FORDHAM
UNIVERSITY
LIBRARIES

QUEUEING THEORY:

A Problem Solving Approach

Not where we stand but in what direction we are moving.

Goethe

Contents

1

Introduction

Waiting in lines seems to be, more often than not, a singularly human pastime. Our daily lives are filled with many waiting-line situations such as those we often encounter in barber shops, cafeteria lines, grocery store checkout counters, highway entrance and exit ramps, parking lots, doctors' offices, bank tellers' windows, toll booths, traffic lights, and so on. Think back for a moment to the many times you have been part of a waiting-line situation in the last hour, the last day, the last week, or the last year. Now think about the many seconds, minutes, and hours you have spent in those lines. Perhaps the information presented in this book may help you enjoy your next waiting-line situation. At the very least, the reasons for such situations may then be better appreciated.

This book cannot provide a complete and current survey of queueing theory, since such a book would require thousands of pages detailing the attendant mathematics and referencing the literature on the subject. This book does, however, present the highlights of the necessarily critical and more typical analysis of queueing systems by presenting the fundamental information required in the application of queueing theory to real-world situations regarding waiting lines.

The mathematical derivations and proofs of the various queueing theory formulas are not, in this author's opinion, as informative or as enlightening to the applications-oriented reader as are the various techniques and methodologies which queueing theory offers. The proofs are, of course, important to those specialists who wish to apply such an approach to solving new problems. However, the reader interested in applying knowledge to a particularly vexing problem is most concerned with the resultant formulas; for this reason, most of the mathematics here is presented in a "cookbook" fashion, that is, rigorous proofs are not given, but rather sufficient mathematics is presented and applied in a result-oriented manner.

It should be noted that queueing theory will not solve all manner of waiting-line problems; it should provide useful and vital information by forecasting or predicting the various characteristics and parameters of the particular waiting line under study. The required decisions made with regard to the waiting-line situation must be made in light of the "answers" provided by the application of various queueing theory formulas and not by the absolute answers given by the application of queueing theory concepts. Queueing theory does not necessarily attempt to develop or provide optimal results as is the usual case for many of the aspects in the operations research arena. Rather, queueing theory merely develops or provides the necessary information which allows the decision maker to study the overall behavior and characteristics of the waiting-line situation under study.

Queueing theory studies waiting-line systems by formulating the appropriate mathematical models of the particular system under study. These models and their operation are then used to derive the necessary performance measurements of the queueing system as a whole. These measurements provide the vital information which will be later utilized in designing an effective waiting-line system. An appropriate balance or trade-off between the various costs and the waiting for the required service are, of course, the bottom line in a waiting-line system. The total cost of service and customer waiting time is the essence of our study of queueing systems.

A queueing system problem is a process where either the arrival rate of the customers into the system or the number of service facilities available for these arriving customers or both of these components are the primary subject of the study. Furthermore, the attendant costs associated with both the waiting time of the customers and the idle or busy times of the servers are under constant review. Many real-world queueing systems can be studied by the appropriate application of the concepts of queueing theory. The list presented previously may be extended to include not only those waiting-line situations in the real world inhabited by humans but waiting line situations involving all manner of customers and servers, for example, airport runways, computerized communications systems, hospital beds, material-handling equipment, repairpersons, ships waiting for an empy pier, trunk lines at a telephone exchange, and so on. The costs associated with providing service for the customers of these queueing systems must be calculated and adjusted in order to minimize the total outlay for the waiting line.

Since queueing theory may be applied effectively to a wide variety of queueing systems, it is not possible to develop or provide a general application that will fit each and every situation. Rather, together with queueing theory this book will attempt to provide a broad picture of the more typical approaches which may be used in determining the overall characteristics and behavior of some of the more specific queueing systems.

Queueing systems may be broken down into three major components:

1. Number of available servers
2. Number of arriving customers
3. Manner in which the servers and customers behave

When a queueing system is viewed in terms of these three major components, the corresponding variables which affect the information provided to the decision maker may be effectively determined, calculated, and presented.

All three components interact in such a manner as to affect the overall behavior of the entire system, that is, the appropriate or the available level of service in terms of time and money is directly affected by the customers, the servers, and the behavior exhibited by the customers and the servers. By the same token, the number of customers demanding service from a queueing system service facility has a tremendous impact on the servers of the system as well as on the customers arriving at the queueing system. Since the three major components interact in such a manner as to impact the entire system, queueing systems must be studied in terms of all the components rather than focusing on any one component.

It should be obvious that the total cost involved in providing a service is normally the major consideration when dealing with queueing systems. Generally, this cost is assigned to the service facility of the system, that is, given a certain number of servers, would it be more or less profitable to increase or decrease the level of service by adding or deleting servers? Obviously, the final decision involves not only the attendant costs of service but the interaction of the customers expecting a particular level of service. If a queueing system has more servers than are necessary to accommodate the level and the number of arriving customers, it would be not only justifiable but reasonable to decrease the server facility to a less costly level of service. However, if customer arrivals demand more service facilities, the question becomes, how much should the service facility be enlarged in order to handle this increase in customer demand?

As with most decision-making processes involving objective and subjective information, there must be a compromise available which affects all considerations with as little impact as possible. The trade-off inherent in queueing systems involves the adjustment of customer waiting times against server idle times. The overall level of service which satisfies the waiting customer and keeps the server busiest is the primary goal of queueing system problem solving. No hard and fast answers are available, unfortunately. However, general information regarding the specific characteristics of a queueing system and the associated behavior of the various elements of the queueing system are available. The major characteristic with which we shall be dealing is time and, of course, time is money.

Credit for the first systematic study of waiting lines is attributed to A.K. Erlang during his investigation of the Danish telephone system in the early 1900s. Other important contributions to queueing theory, especially from the mathematical point of view, have been made by Cox and Smith, Kendall, Lindley, Saaty, and Takacs. Recent studies by Allen, Buzen, Chang, and Kleinrock have added much to the earlier works of queueing theorists. A complete and current reference listing of those who have made major advances in queueing theory would be, to say the least, a major undertaking. The Bibliography lists, at best, a partial list of the available literature. The serious reader is directed to the references as a starting point to further study of queueing theory. The references also include literature concerned with the study of probability theory. As will soon be seen, probability theory forms the basis of many of the concepts of queueing theory.

The material in this book falls into two major areas: fundamentals of probability theory and queueing theory applications. The choice of this particular outline was necessary because the concepts utilized in probability theory play an important role in the subsequent development and presentation of the concepts of queueing theory.

Probability theory is primarily concerned with the study of random experiments, that is, it is a mathematical study of occurrences which, prior to their occurrence, are unknown to the observer. The early work of probability theorists involved the investigation of various games of chance, particularly card and dice-tossing games. Modern probability theorists apply their craft in almost every discipline. Biology, chemistry, and physics use probability theory as the basis for much of the results of their study. All fields of engineering use probability theory, to some extent, in the forecasting and calculating of the various attributes of the particular items under study. Business, education, and sociology employ concepts from probability theory in predicting or modeling the various subjects of behavior. Our use of probability theory will focus on its application in the particular operations research area known as queueing theory.

Queueing theory concerns itself with the mathematical development of models which will forecast the behavior of waiting-line systems in which customers arrive and demand service of the system. Because these customers demand service in terms of some law of probability, queueing theory has been considered within the realm of probability theory, especially within the structure of stochastic or random processes.

Queueing theory owes much of its early development to this past reliance on the concepts provided by probability theory. Furthermore, a solid but fundamental knowledge of probability theory is usually required in order to appreciate fully the intricacies of the concepts which lie at the heart of queueing theory. The aspects of probability theory presented in this book should provide this fundamental background necessary to our later presentation regarding queueing theory concepts. The serious reader should refer to the literature listed in the Bibliography for the in-depth material required by the specialists in the queueing theory field.

The basics of probability theory required in this book are usually covered in a first-level college course in probability theory. The reader with such a background may use the initial chapters of this book simply as a review. A thorough understanding of fundamental probability theory may be reason enough to skip over this section entirely. For those readers without this prerequisite knowledge of probability theory, the initial chapters will provide the necessary basic information required to appreciate later queueing theory concepts; the references dealing with probability theory will provide a wealth of supportive and additional information with regard to the concepts forming the basis of probability theory.

The programmed routines presented later were written as BASIC-language coding, since this particular programming language is readily available on most minicomputers and microcomputers. The availability of a computing system which will accept these routines as written will greatly enhance the overall understanding of the terminology and mechanics of queueing systems. In particular, the programmed routines presented in the upcoming chapters were run successfully on a TRS-80™ (Level II BASIC with 16K memory). Readers with access to this particular microcomputing system will experience no problems in running the code as it appears in the text. Readers who have access to another dialect of BASIC should experience little difficulty in obtaining a converted verion of the coding, since none of the significant features of the TRS-80™ Level II BASIC are implemented in the programmed routines.

2 Probability Theory Concepts

The basic concepts and terminology of probability theory are a necessary requirement in order to appreciate fully queueing theory concepts and terminology. For this reason, we shall introduce those notions of probability theory which are required as our steppingstones into the field of queueing theory. In particular, the following ideas will be presented as an initial guide:

1. Fundamental concept of randomness
2. Experiments, sample space, and events
3. Relative frequency
4. Probability measure
5. Random variables
6. Probability density function
7. Cumulative distribution function
8. Mean and variance
9. Transformation methods

These concepts should be familiar to those readers who have had a first-level college course in probability theory. Readers interested in a basic review of probability theory concepts will get such an understanding from this chapter. Readers primarily interested in an in-depth study of probability theory, however, are directed to the Bibliography. Finally, readers with a firm grasp of probability theory may find this chapter rewarding in terms of a quick review. The primary intent, at this point, is merely to introduce, in as painless a manner as possible, the fundamental concepts of probability theory required for our further study of queueing systems.

Randomness

The fundamental notion of probability theory involves the description of events which occur in a random manner. The key word in this initial definition is *random*. If we were to take the time to look up a dictionary definition of the word random, we would (in all probability) find it described in terms of aimlessness, haphazardness, or irregularity. The fact is, however, that a random event, for our purposes, has none of these qualities. The concept of regularity forms the cornerstone of randomness as it applies to the various tenets of probability theory.

As an example of what randomness involves, let us perform a simple experiment. The manner in which this experiment will be performed and the results obtained should give a better appreciation of randomness as it applies to probability theory. The experiment concerns the tossing of a six-sided die and the tabulating of the values which appear face-up after the toss. We must also assume that the die is fair, that is, we expect any value to appear face-up with the same regularity as any other value of the die. In other words, a six should appear face-up one out of six tosses, as should a one, or a three, and so on. This assumption of fairness will form the basis of our description of the word random.

Probability theory was first developed and used in the description of various game-playing situations. For this reason, our initial introduction will also center about a gaming situation. The die-tossing game is a rather simple but effective experiment on which to base the definition of randomness and will be the vehicle for which most of the concepts of probability theory will be based. Let us continue with our tossing of the die.

Rather than spend time physically tossing a six-sided die and tabulating the results of the face-up value after each toss, let us use a computer to accomplish this task. The programmed routine listed in Figure 2-1 and the accompanying remarks should enable the user of this routine to understand the logic of the coding. Running this program will produce output similar to that listed in Figure 2-2. Let us turn our attention to these results.

```
1000 REM 6-SIDED DIE-TOSS
1010 REM VARIABLE-NAME     DESCRIPTION
1020 REM         D         ARRAY WHICH HOLDS THE 6 VALUES
                               OF THE DIE
1030 REM         F         CUMULATIVE DISTRIBUTION FUNCTION
1040 REM         J         GENERAL LOOP COUNTER
1050 REM         N         GENERATED PSEUDORANDOM NUMBER
1060 REM         T         NUMBER OF TOSSES
1070 DIM D(6)
1080 FOR J = 1 TO 6
1090   D(J) = 0
1100 NEXT J
1110 REM   COMMENCE TOSSING THE DIE (T) TIMES
```

FIGURE 2-1: Programmed routine die-toss

```
1120 INPUT "NUMBER OF TOSSES";T
1130 FOR J = 1 TO T
1140   N = INT(RND(6))
1150   D(N) = D(N) + 1
1160 NEXT J
1170 PRINT "VALUE       ABSOLUTE       RELATIVE";
1180 PRINT "       CUMULATIVE"
1190 PRINT "            FREQUENCY     FREQUENCY"
1200 PRINT "   DISTRIBUTION FUNCTION"
1210 PRINT "             (F)        F(T)"
1220 PRINT "         F(X)"
1230 FOR J = 1 TO 6
1240   F = 0.0
1250   PRINT J, D(J), D(J)/T,
1260   F = F + ( D(J) / T)
1270   PRINT F
1280 NEXT J
1290 GOTO 1080
1300 END
```

FIGURE 2-1: (Continued)

NUMBER OF TOSSES? 10

VALUE	ABSOLUTE FREQUENCY (F)	RELATIVE FREQUENCY (F/T)	CUMULATIVE DISTRIBUTION FUNCTION F(X)
1	2	.20000	.20000
2	0	.00000	.20000
3	0	.00000	.20000
4	4	.40000	.60000
5	2	.20000	.80000
6	2	.20000	1.00000

NUMBER OF TOSSES? 100

VALUE	ABSOLUTE FREQUENCY (F)	RELATIVE FREQUENCY (F/T)	CUMULATIVE DISTRIBUTION FUNCTION F(X)
1	22	.22000	.22000
2	15	.15000	.37000
3	17	.17000	.54000
4	17	.17000	.71000
5	12	.12000	.83000
6	17	.17000	1.00000

NUMBER OF TOSSES? 1000

FIGURE 2-2: Typical output from programmed routine die-toss

VALUE	ABSOLUTE FREQUENCY (F)	RELATIVE FREQUENCY (F/T)	CUMULATIVE DISTRIBUTION FUNCTION F(X)
1	161	.16100	.16100
2	137	.13700	.29800
3	160	.16000	.45800
4	176	.17600	.63400
5	186	.18600	.82000
6	180	.18000	1.00000

NUMBER OF TOSSES? 10000

VALUE	ABSOLUTE FREQUENCY (F)	RELATIVE FREQUENCY (F/T)	CUMULATIVE DISTRIBUTION FUNCTION F(X)
1	1684	.16840	.16840
2	1699	.16990	.33830
3	1684	.16840	.50670
4	1691	.16910	.67580
5	1576	.15760	.83340
6	1666	.16660	1.00000

NUMBER OF TOSSES? 100000

VALUE	ABSOLUTE FREQUENCY (F)	RELATIVE FREQUENCY (F/T)	CUMULATIVE DISTRIBUTION FUNCTION F(X)
1	16688	.16688	.16688
2	16703	.16703	.33391
3	16617	.16617	.50008
4	16587	.16587	.66595
5	16630	.16630	.83225
6	16795	.16795	1.00000

FIGURE 2-2: (Continued)

When we toss the die, we expect each possible value of the die to appear face-up one out of six tosses (on the average, that is). Mathematically, the probability of occurrence of any one value appearing face-up after any toss of the die would be equal to 0.166 . . . , or one out of six. If, after a number of tosses, we noticed that one particular value was appearing more or less often than any other value, we would begin to question the honesty of the die. When we toss the die ten times, the programmed routine does not yield the results which correspond to the theoretical value of one-sixth, however. As the number of tosses or trials increases, from ten through 100,000 trials, the resulting output shows that the theoretical value of 0.166 . . . is approached.

A basic tenet of probability theory involves the fact that a large number of trials must be run before any theoretical value of the experiment is approached. In fact, the law of large numbers states symbolically that

$$\lim_{n \to \infty} \frac{n_A}{n} = P(A)$$

As the number of trials of an experiment (n) approaches infinity (i.e., becomes very large), the ratio of occurrences of the event A (n_A) to the number of outcomes in the experiment (n) will equal the theoretical probability of the event A occurring, that is, $P(A)$. Our experiment did not seek to prove that tossing a fair six-sided die would give each value appearing face-up one out of every six trials. Rather, it did prove that, on the average (that is, after a large number of trials), the actual results approached the theoretical value expected of the experiment.

A number of important concepts of probability theory may now be introduced. The following terms will be used later in our study of both probability theory and queueing theory:

1. Experiment
2. Sample space
3. Events
4. Relative frequency
5. Probability measure

They are defined below with regard to our die-tossing experiment as presented in figures 2-1 and 2-2.

An *experiment*, with regard to probability theory, is the phenomenon under study. In our die-tossing example, the experiment consisted of the rules and the regulations which we intially set forth. We tossed a fair six-sided die and tabulated the face-up values. Each execution of the experiment was carried out under similar conditions, that is, we always used the same die (a pseudorandom number generator), we always tabulated the face-up value of the tossed die, and we always represented the probability of occurrence of each face-up value as a ratio with regard to the total number of tosses made during the experiment.

The *sample space* of a given experiment, denoted by the capital letter S, consists of the set of results or sample points corresponding to each individual and possible outcome of the experiment. The sample space of our die-tossing experiment consists of the set of positive integers one through six, that is, each possible value for each face of our original six-sided die. A sample point or an element of the sample space S is, therefore, one particular and possible outcome. Mathematically, the sample space of our die-tossing experiment would be expressed symbolically as

$$S = (1, 2, 3, 4, 5, 6)$$

or, the set of positive integers one through six.

When any particular sample point or element occurs during the execution of a given experiment, we call this occurrence an *event* or an outcome of the experiment. When the face value three appeared face-up, this occurrence corresponded to one such event in the experiment. The appearance of a five, for example, would constitute yet another event of the experiment.

Mathematically, an event is a set of outcomes or a subset of the given sample space S. The mathematical definition of our experiment could be written as

$$S = (1, 2, 3, 4, 5, 6)$$

that is, the sample space S which consists of the elements 1, 2, 3, 4, 5, and 6. An additional definitive notion about our experiment is that the sample space which we are using is a finite sample space, that is, the number of elements presented in the given sample space are countable and finite. This particular concept is not very important right now; when the time comes to introduce an infinite sample space we shall examine more thoroughly the notion of finite and infinite sample spaces.

Referring to the output produced by our die-tossing routine (Figure 2-2), let us now look at the Absolute Frequency (f) column which lists the number of occurrences of each possible outcome. In particular, this column is called the absolute frequency of the event E where E = 1, 2, 3, 4, 5, 6. When the value which corresponds to the number of trials (t) is divided into the absolute frequency (f) for each event (E), the resulting quotient is called the relative frequency for that event in a given experiment. The relative frequencies for each possible event in our die-tossing experiment are the values listed under the column heading Relative Frequency (f/t).

As can be seen, as the number of trials (t) increases, the values attained by the relative frequencies for each event stabilizes to the theoretical value of 0.166 ..., or the value which should be attained if the die is fair. As these relative frequencies approach the theoretical value, the fundamental basis of probability theory begins to come to light. This stabilizing of the relative frequencies of each event of our given experiement, determined by random occurrences of possible events, lends a definitive degree of regularity to the word random.

A *probability measure* (P) is a one-to-one mapping or reassigning of possible experimental events defined on a sample space S onto the set of real numbers R. In mathematical notation, P corresponds to the value of the relative frequency of an event (E), and $P(E)$ denotes the set of real number associated with the particular event E. The following requirements must be satisfied in order for this mapping of events onto the real numbers to occur.

For any event A,

$$0 \leqslant P(A) \leqslant 1$$

that is, the relative frequency of any given event must have a real value greater than or equal to zero and less than or equal to one. In our experiment, each event approached a theoretical value of 0.1666 ... , or one-sixth.

The relative frequency of our sample space S, which consists of the set of positive integers one through six, must total one. In order to accumulate the values of each relative frequency, it is necessary to determine the relative frequency values of each event as a probability of occurrence P_i, where $i = 1, 2, 3, 4, 5, 6$. That is,

$$\sum_{i=1}^{6} p_i = 1$$

The experiment or process we just examined forms the basis of what is referred to in probability theory nomenclature as a stochastic or a random process. In order to simulate or model a real-world situation or experiment, it is often desirable to use the concept of a random process rather than the actual real-world situation. Since we have defined random as a statistically regular system of occurrences and since most real-world situations can be accommodated by a mathematically random model, the notion of a random process takes on much importance. The role of a random process will be further displayed when we begin to look at some of the queueing or waiting-line problems associated with our everyday lives.

In summary, we shall use the word random to indicate a particular degree of regularity with regard to the determining of the probability of occurrence of certain events during the execution of a given experiment. We usually know the range of the random events; however, the exact sequence of these events is not (usually) known beforehand. If, as in the die-tossing experiment, we knew that a six would appear face-up after a three fell face-up, the element of randomness, as well as the element of mystery, would not be a part of the experiment. The successful gambler or game-player would, of course, like to know the exact sequence of events in the game being played. However, probability theory only affords the range of values which the events may take on during the execution of the game or experiment and not the sequence of these events.

Random Variables

As we discussed earlier, the word random implies an equal probability of occurrence, not haphazardness as is usually the case. The accompanying definitions of the terminology used in probability theory must be understood in terms of their relationship to probability theory and not in terms of their everyday and, quite often, misused manner.

The stochastic or random process—which forms the foundation of probability theory—is used in a variety of ways to model or simulate a real-world situation or experiment. For example, an epidemiologist might study the occurrence, the distribution, the severity, and other important parameters of an epidemic in a given population by conducting mathematical experiments upon a hypothetical population rather than on the real-world population suffering from this epidemic. On the other hand, gamblers may (unknowingly perhaps) refer to random processes during their quest for the elusive big winner. The particular characteristic of a random process which makes it so appealing to the experimenter is its quality of randomness. The ability of the process to assign certain probabilities to events occurring during the execution of the experiment in such a manner as to define totally the range of the events lends a degree of regularity or stabilization to the experiment.

The values attained by a random process for each of its possible outcomes is known as a *random variable*. Although the exact sequence of the values attained by this random variable are unknown, the range of these values is, in fact, known beforehand. The range of possible values which our die-tossing experiment was able to assume was the set of positive integers one through six. The probability of occurrence of each of these values was also known beforehand, namely, one-sixth. A random variable, denoted by the capital letter X, is defined as a nondeterministic variable which may be described completely in probabilistic terms.

Random variables are usually represented in terms of the probability distribution function which describes the probability of the various values taken by the random variable during a given number of trials or experimental events. In mathematical terminology, a random variable is a value which depends upon the outcome of a particular experiment and no other factors. The value attained by the random variable is determined by the mapping or reassigning of the sample points from a given sample space onto the set of real numbers. In other words, a random variable is a real-valued variable defined in terms of the probability function which maps the sample points of the given experiment onto the set of real numbers.

Random variables may be classified as being either discrete random variables or as continuous random variables. Since the values which the random variable may assume lie on the real number line, this particular characterization should not come as a surprise. Both the discrete random variable and the continuous random variable will be presented in terms of their probability theory nomenclature.

Discrete Random Variables

A discrete random variable may assume only a finite number or, at most, a countably infinite number, of values during the course of an experiment. If a random variable X can take on the different values x_i (where $i = 1, 2, 3, \ldots, n$) and the probability of the value x_i is defined as p_i (i.e., the set of values p_i where $i = 1, 2, 3, \ldots, n$), then it can be said that the random variable X has a discrete probability function. Our die-tossing experiment is an example of a random variable which may be defined in terms of a discrete probability function. The values x_i taken by the random variable X in our experiment correspond to the set of positive integers one through six, and the probability (p_i) of each value (x_i) is equal to one-sixth.

Since the random variable must take one of the six possible values p_i, it follows that the summation of all possible p_i's must be equal to one. Mathematically, this summation of the probability values may be stated as:

$$\sum_{i=1}^{n} p_i = 1$$

In fact, if we were to accumulate the p_i values for our die-tossing experiment, we would find that, since all p_i's are equal to 0.1666 . . . and there are six possible values which x_i may assume during the course of the experiment, the above summation does hold true.

In summary, if the set of random variables taken is of the form

$$x_i \quad \text{where } i = 1, 2, \ldots, n \tag{2-1}$$

and the probability of occurrence of the value x_i is of the form

$$p_i \quad \text{where } i = 1, 2, \ldots, n \tag{2-2}$$

and the following summation holds true,

$$\sum_{i=1}^{n} p_i = 1 \tag{2-3}$$

then the random process described in equations 2-1 through 2-3 has a discrete probability function. In other words, the values taken by the random variable X may assume only a finite or, at most, a countably infinite, number of values.

Discrete Probability Function

Assume that the random variable X on the sample space S, where the mapping of X is onto the finite set of real numbers, is described by

$$X(S) = (x_1, x_2, \ldots, x_n)$$

The probability of x_i is defined as

$$P(X = x_i)$$

and the function of x_i is written as $f(x_i)$. The distribution or the probability distribution function of the random variable X is, therefore, given by

$$f(x_1) = x_1$$
$$f(x_2) = x_2$$

and so on to

$$f(x_n) = x_n$$

where

$$\sum_{i=1}^{n} f(x_i) = 1$$

In our die-tossing experiment, the probability distribution function of the random variable X could be graphed as shown in Figure 2-3. The random variable X, therefore, has the following mapping onto the set of real numbers:

$$X(S) = (1, 2, 3, 4, 5, 6)$$

The distribution f of the random variable X is:

$$f(1) = P(X = 1) = 1/6$$
$$f(2) = P(X = 2) = 1/6$$
$$f(3) = P(X = 3) = 1/6$$
$$f(4) = P(X = 4) = 1/6$$
$$f(5) = P(X = 5) = 1/6$$
$$f(6) = P(X = 6) = 1/6$$

Figure 2-3 illustrates graphically the distribution we just examined. Our being able to define a random variable as a random process as well as to represent the process graphically will assist us to a great degree later in our examination of the various distributions employed in queueing systems.

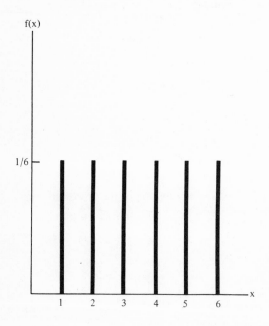

FIGURE 2-3: Probability distribution function of the die-tossing experiment

Continuous Random Variables

A continuous random variable, on the other hand, can take all the values in a finite set or all the values in an infinite set of the real number line. In other words, if a random variable X is of the set of real numbers $-\infty < x_i < \infty$ (where $i = 1, 2, 3, \ldots, n$), the random variable X is said to lie in a continuum of the real number line. The discrete random variable was limited to the values which it was able to assume to a particular finite set of values; the continuous random variable is able to assume all the possible values within a particular subinterval of the same real number line.

An example of a continuous random variable would be the instantaneous speed of the tossed die at some point in time after it leaves the tosser's hand. A corresponding example of a discrete random variable would, of course, be the particular discrete value the die assumed after it came to rest face-up.

Continuous Probability Function

Mathematically, a continuous probability function, denoted by $f(x)$, for all real-valued x's must be greater than or equal to zero. In addition, the following must hold true:

$$\int_{\infty}^{-\infty} f(x)dx = 1 \quad \text{where } f(x) = 0 \tag{2-4}$$

If the range of the continuous random variable X lies in the subinterval (a,b) and the probability that X assumes a value in any subinterval (a,b), then the area under the curve over this particular subinterval is given by:

$$P(a \leqslant X \leqslant b) = \int_{b}^{a} f(x)dx \quad \text{where } f(x) = 0$$

The function of x, that is, $f(x)$, is called a continuous probability density function. The area under the curve bounded by $x = a$ and $x = b$ corresponds to the probability that the continuous random variable X assumes a value x in the real-valued subinterval (a,b). Since the values a and b lie on the real number line, they are considered to be an ordered pair of real numbers and are usually written as (a,b).

Probability Density Function

Being able to represent mathematically the characteristics of a random variable, be it a discrete random variable or a continuous random variable, is a necessary first step in the further determination of the requirements of probability theory as it applies to queueing theory applications. The next step is to represent these random variables as a function rather than as the values which they may assume during the execution of a given experiment.

The probability density function is a convenient form which allows us to exhibit the probabilities which correspond to a random variable. Since we are most interested in knowing the probability of occurrence of any particular value which a random variable X may assume, the probability density function becomes the vehicle with which we may identify these particular values.

Another descriptive form of a random variable is that of a cumulative distribution function. This function defines the probability that the value x_i taken by the random variable X is less than or equal to some given value x. If $f(x)$ is the probability density function, then

$$F(x) = \int_{-\infty}^{x} f(x)dx \qquad (2\text{-}5)$$

where $0 \leqslant F(x) \leqslant 1$, and the probability of x falling in the range x_1 to x_2 (or a to b) where $x_1 \leqslant x \leqslant x_2$ is given by:

$$F(x_2) - F(x_1)$$

A further examination of both the probability density function and the cumulative distribution function is necessary, since both of these probability functions play an important role in queueing theory as well as in probability theory.

As mentioned previously, the particular value x, which a random variable X may assume during the execution of a given experiment, is not as important as is the value of the probability of occurrence of that random variable's value. A gambler may be quite interested in knowing the value thrown face-up for a six-sided die; however, in order to appreciate fully and to play our die-tossing game intelligently, it is of paramount importance to know the probabilities of the values which may be thrown. Therefore, a probability density function is used as the primary mathematical tool in the die-tossing. The probability density function of the six-sided die-tossing experiment may be written as

$$f(x) = 1/6$$

for $x = 1, 2, 3, 4, 5, 6$, where x represents the event that a 1, 2, 3, 4, 5, or 6 appeared face-up after the die was tossed.

The probability density function for this discrete random experiment can be determined from the following description. The experiment is, of course, random, and, in fact, it is a discrete random experiment, since the values taken by the random variable X lie within a finite set of real numbers, namely, the set of positive integers one through six. The sample space S of the experiment contains six possible outcomes, that is, the uppermost face having a value of one, the uppermost face of the tossed die having a value of two, and so on to the uppermost face of the die having a value of six.

Suppose we toss our fair six-sided die 100 times and tabulate the number of occurrences for each face-up value, that is, we tabulate the absolute frequency (f) of each possible outcome. Refer again to Figure 2-2 for these values. If the die is fair, we expect the probabilities of these various possible outcomes to be equal according to our previous definition of the probability density function given by $f(x) = 1/6$.

The Relative Frequency column of Figure 2-2 lists the actual observed frequencies in each series of tosses or trials. The theoretical value of one-sixth becomes our probability density function value. Figure 2-3 is an illustration of these theoretical values for our probability density function.

Cumulative Distribution Function

The cumulative distribution function, denoted by $F(x)$, of our die-tossing experiment is also tabulated in Figure 2-2. The column of data labeled $F(x)$ represents the probability that the random variable X takes on the value which is less than or equal to some value x_i. The particular value of $F(x)$ for any given value of x_i can be determined by accumulating the values for $f(x)$ for all values of the random variable less than or equal to the value being considered. In our experiment (for 100 trials),

$$F(1) = f(1) = 0.22$$
$$F(2) = f(1) + f(2) = 0.22 + 0.15 = 0.37$$
$$F(3) = f(1) + f(2) + f(3) = 0.22 + 0.15 + 0.17 = 0.54$$

and so on. It should be noted that if parameter a is the smallest value of x while parameter b is the largest value of x, then the following condition holds true:

$$F(x) = 0 \quad \text{for all } x \leqslant a$$
$$F(x) = 1 \quad \text{for all } x \geqslant b$$

The graph of the cumulative distribution function $F(x)$ is, for our die-tossing experiment, represented as a step function where the x_i's are plotted on the abscissa and the corresponding $F(x)$ are plotted on the ordinate in a typical Cartesian coordinate system. The cumulative distribution function of our discrete random experiment is illustrated in Figure 2-4.

In summary, the cumulative distribution function of a random variable defines the probability that the observed value is less than or equal to some value x. If $F(x)$ is the cumulative distribution function of a continuous random variable X, then

$$F(x) = \int_{-\infty}^{x} f(x)dx$$

where $f(x)$ is defined as the probability density function of the continuous random variable X. From its definition, $F(x)$ is a positive real number with a range of zero to one, and the probability of some x falling in the range (a,b) is defined as $F(b) - F(a)$.

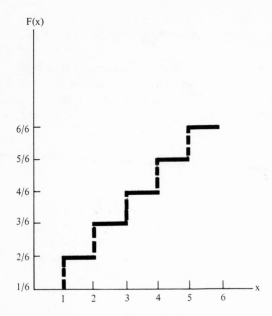

FIGURE 2-4: *Cumulative distribution function of the die-tossing experiment*

If the cumulative distribution function of a discrete random variable X has probability p, then the sum of all x_i, such that x_i is less than or equal to x, is the cumulative distribution function or

$$F(x) = \sum_{x_i \leqslant x} p(x_i)$$

An everyday example of a discrete random variable X involves the cumulative distribution function $F(x)$ which is illustrated in Figure 2-5. The curve of this random variable is a step function. As can be seen from the graph, this particular discrete function is quite representative of the postage rates charged for First Class mail (as of 1979), that is, 15 cents for the first ounce, 30 cents for a parcel of mail larger than two ounces and less than three ounces, and so on.

Another example which illustrates the respective graphs of a continuous random variable X is illustrated in figures 2-6 and 2-7 for the following probability density function and corresponding cumulative distribution function:

$$f(x) = \begin{cases} x/2 & \text{for } 0 \leqslant x \leqslant 2 \\ \\ 0 & \text{elsewhere} \end{cases}$$

$$F(x) = \begin{cases} 0 & \text{for } x < 0 \\ x^2/4 & \text{for } 0 \leqslant x \leqslant 2 \\ 1 & \text{for } x > 2 \end{cases}$$

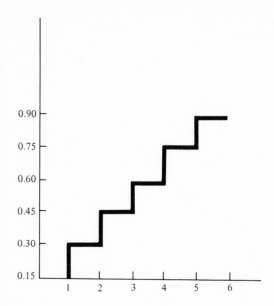

FIGURE 2-5: *Cumulative distribution function of postal rates*

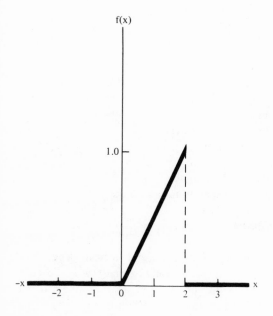

FIGURE 2-6: *Probability density function for f(x) = x/2 for 0 ⩽ x ⩽ 2*

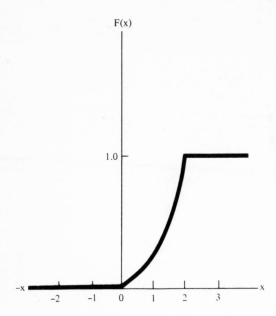

FIGURE 2-7: Cumulative distribution function for F(x) = $x^2/4$

Further Topics

Various measures associated with statistical distributions can provide a deeper insight into some of the more important characteristics of the particular distribution under study. The most important of these additional measurements use the moments of a distribution in determining the effective values for these measures. The importance of moments of a random variable X lies in the information which these parameters convey about the distribution of the given random variable. The expected value or the mean of the distribution clearly contains sufficient information related to the average value of a given random variable. The second central moment of a distribution conveys numerical information regarding the variability of the random variable about its mean value; in other words, the variance illustrates the mathematical spread of the given distribution about its previously calculated mean or average value.

The first moment of a distribution is called the mean and is usually denoted by the Greek symbol mu (μ). As with the arithmetic mean or average value, the mean or expected value of a particular probability distribution is used to indicate the central tendency or location of the given probability distribution.

Moments may also be taken which relate the variance or the measure of dispersion of a given probability distribution. In addition, the standard deviation may be calculated for the given distribution as a further statistical measurement of the probability distribution and its related random variable. The calculations of these moments involve the parameters of the probability distribution which we had developed earlier.

The mean or the expected value (μ) of a discrete random variable X is calculated by multiplying each of the possible values of the random variable by the corresponding values for the probability of occurrence of each random variable value. Forming the sum of all these products yields the expected value or mean. The mean value need not, of course, occur among the individual values of the discrete random variable, just as the mean value of a series of numbers need not necessarily occur among those values.

In general, the mean or expected value of a discrete random variable X which takes the discrete values $x_1, x_2, \ldots, x_n$ and the corresponding probabilities of occurrence $p_1, p_2, \ldots, p_n$ is given by:

$$\sum_{i=1}^{n} x_i p_i \tag{2-6}$$

The mean (μ) of a discrete random variable X is simply the weighted average of all possible values of the random variable (x_i's) weighted by the random variable's probabilities (p_i's).

Returning to our die-tossing experiment, we have the following calculation which yields the mean or expected value:

$$\mu = \sum_{i=1}^{6} x_i p_i$$
$$\mu = x_1 p_1 + x_2 p_2 + x_3 p_3 + x_4 p_4 + x_5 p_5 + x_6 p_6$$
$$\mu = 1(1/6) + 2(1/6) + 3(1/6) + 4(1/6) + 5(1/6) + 6(1/6)$$
$$\mu = 3.5$$

The average value of our discrete random variable which defines our die-tossing experiment is equal to a value of 3.5. It should be noted that this calculated value does not appear as any of the possible face-up values of our given discrete random variable.

The variance or second central moment of a discrete random variable X is calculated by multiplying the square of each deviation from the mean value by the corresponding probability and forming the sum of these products:

$$\sum_{i=1}^{n} (x_i - \mu)^2 p_i \tag{2-7}$$

Our die-tossing experiment had a calculated mean of 3.5, and the following probability function defines this discrete random variable:

x_i	1	2	3	4	5	6
p_i	1/6	1/6	1/6	1/6	1/6	1/6

Hence, the variance is calculated by:

$$\sigma^2 = (1 - 3.5)^2 1/6 + (2 - 3.5)^2 1/6 + \ldots + (6 - 3.5)^2 1/6$$
$$\sigma^2 = 2.92 \text{ (approximately)}$$

For a continuous random variable X, the mean (μ) is calculated by multiplying the value of the generated random variate (x) by the probability density function $f(x)$ and solving this product by integrating it over the range $-\infty$ to ∞, that is:

$$\mu = \int_{-\infty}^{\infty} x \; f(x) \; dx \tag{2-8}$$

The variance (σ^2) of a continuous random variable X may be calculated by squaring the derivations from the mean of each x_i and multiplying this derivation by the appropriate $f(x_i)$, that is, the probability density function. Each of these products are summed over the range i to n:

$$\sigma^2 = \sum_{i=1}^{n} (x_i - \mu)^2 f(x_i) \tag{2-9}$$

The variance of a continuous random variable X is the second central moment of the given random variable. In addition, the root mean square of the variance is, of course, the standard deviation:

$$\sigma = (\sigma^2)^{1/2} \tag{2-10}$$

In summary, the mean or the expected value of a random variable X yields a typical or average value for X. The variance reflects the relative spread of the values of a random variable with respect to the mean value of a given random variable X. The standard deviation of a random variable X is a measure of the tendency of the values of X to cluster about the mean value of X. While other moments of a random variable may be important in certain other situations, the mean and the variance as well as the standard deviation are considered to be the most important and are, therefore, the moments associated with the calculations of other parameters of a particular probability distribution.

Generation of Random Variates

In order to obtain a representative set of random variates (x_i's) for a particular probability distribution, it is necessary to use an operation which involves some sort of numerical generating technique. We may simply employ a table of generated random numbers which can be found in most probability theory textbooks or use some other means at our disposal. Our primary concern, however, is to be able to generate the random variates which correspond to the given probability distribution and not to any statistical distribution. We may use the standard library function found in most BASIC-language interpreters which is known as RND. In fact, this particular function will be used as the basis for much of the work we shall be doing from now on. The values obtained from this library function will not, however, correspond exactly to the form of the random variates of a given probability distribution without some additional mathematical manipulation. The primary reason for this required manipulation of the resultant random values is that the RND function generates pseudorandom numbers which are totally unacceptable in their original form and dispersion for the variety of probability distributions which we will eventually encounter during our examination of queueing systems. We must have at hand a method which can easily and effectively convert or transform the values given by the RND function into their appropriate random variates for the particular probability distribution which we are studying.

Generally, the modeling of a particular random process involves the replacement of the actual statistical elements or values of a given process by an appropriate theoretical value while governing this replacement by the particular probability distribution which mathematically describes the given random process. The inverse transformation method is but one of the many methods commonly utilized to effect this replacement of actual to theoretical values in an appropriate fashion. Since the inverse transformation method is, perhaps, the simplest technique at our disposal, we shall use it to generate the necessary random variates. The rejection method, the comparison method and a variety of other mathematical devices are available which accomplish similar results; these techniques take advantage of some of the abilities of the device which will generate the required random variates. Maisel[1] and Naylor[2] treat the application of the rejection method and the composition method as well as give a readable description of the inverse transformation method.

Let us now briefly describe the operation of the inverse transformation method as it applies to our examination of random variate generation. Initially, we must assume that we have a certain probability density function, $f(x)$, and the means with which to convert it into its corresponding cumulative distribution function, $F(x)$. Our primary concern is to be able to generate the required random variates (x_i's) from this given probability distribution in order to accommodate an appropriate modeling of the given distribution in mathematical terms.

[1] H. Maisel, and G. Gnugnoli, *Simulation of Discrete Stochastic Systems* (Chicago: Science Research Associates, 1972), pp. 42–54.
[2] T.H. Naylor et al. *Computer Simulation Techniques* (New York: Wiley, 1966), pp. 70–77.

From the given probability density function, $f(x)$, we determine the corresponding cumulative distribution function, $F(x)$. In addition, $F(x)$ must be limited to the range of values from zero to one in order to accommodate an easy application of the RND library function of our BASIC-language interpreter. Under most interpreters, coding this function as RND (0.0) will yield pseudorandom numbers in the range of real numbers from zero to one. Therefore, $F(x)$ must be limited to this range in order to easily code the necessary call to the RND function. The necessary random variates will be generated through the RND function and transformed via the inverse transformation method to the subinterval defined for the particular probability distribution under examination.

Since $F(x)$ is initially defined over the range zero to one, we may make the assumption

$$F(x) = r$$

where r is the generated pseudorandom number from the RND function. Since the range of the available pseudorandom numbers lies in the interval zero to one and the range of the given cumulative distribution function lies in the same subinterval of the real number line, this assumption may be made without further manipulation. An additional requirement of $F(x)$ is that it must be an increasing function in the given subinterval (a,b) as the value of x increases. In other words, the cumulative distribution function, $F(x)$, must be able to assign a unique pseudorandom number (r) to one and only one value of x, that is, for each r_i (where $i = 1, 2, \ldots, n$) there exists a unique x_i (where $i = 1, 2, \ldots, n$).

Mathematically, if $F(x) = r$, then, by the inverse transformation method, $x_i = F^{-1}(r_i)$ where $i = 1, 2, \ldots, n$. The term $F^{-1}(r)$ is a mapping of the value r on the subinterval zero to one of the real number line.

An example of the application of the inverse transformation method to a common probability distribution function is in order. Initially, we define the probability density function of the example probability distribution. For our purposes, this example will use the uniform probability distribution. The given probability density function for the uniform probability distribution is given as

$$f(x) = 1/(b - a)$$

where the ordered pair of real numbers (a,b) are related by $a \leqslant x \leqslant b$. The corresponding cumulative distribution function for the uniform probability distribution is given by

$$F(x) = (x - a)/(b - a)$$

where $a \leqslant x \leqslant b$. Furthermore, $F(x)$ is defined as being limited to the range of real numbers zero to one in order to accommodate the application of the RND library function later in this operation. The graphs associated with the probability density function and the cumulative distribution function for the uniform probability distribution are illustrated in figures 3-1 and 3-2, respectively.

We now would like to take the inverse transformation of the cumulative distribution function $F(x)$ by the following:

$$F^{-1}(r) \quad \text{where } x = F^{-1}(r)$$

In order to accomplish this transformation, we must be able to generate a uniform random variate which is defined over the range zero to one and apply that generated random value (r) to $x = F^{-1}(r)$. For each such generated random value (r), a unique value of x must be associated. This one-to-one association is, indeed, possible due to the fact that the defined cumulative distribution function for the uniform probability distribution is an increasing function over the interval zero to one as the value of x increases; see Figure 2-8. Therefore, if we were to generate the random variate r_1, it would map onto the associated value x_1. Likewise, the random variate r_2 would map onto x_2 and so on.

A programmed routine which may be used to generate the appropriate random variates for a uniform probability distribution is described in greater detail in the section dealing with the uniform probability distribution in chapter 3.

A second example which employs the inverse transformation method to generate the required random variates has the probability density function

$$f(x) = 2x$$

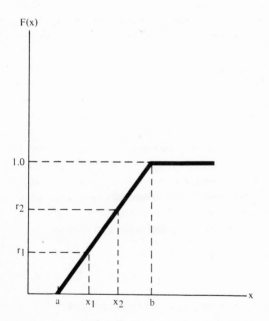

FIGURE 2-8: *Graphical representation of unique x_i for each r_i*

where x is defined over the interval zero to one. The corresponding cumulative distribution function is given as

$$F(x) = x^2$$

where, again, x is limited to the interval zero to one.

The inverse transformation of $F(x)$ is given by

$$x = F^{-1}(r) = (r)^{1/2}$$

where the range of x is mapped onto the random variate r. In other words, values of x_i having a probability density function equal to $2x$ may be generated easily by calculating the square root of the generated random value r.

Our ability to generate the appropriate random numbers or to instruct a computer to accomplish the generation of appropriately distributed random numbers is the subject of much of the literature devoted to this aspect of probability theory. Maisel offers a descriptive outline and a FORTRAN routine which present some of the more appropriate techniques used in the generation of random numbers. For our purposes, when a random number is to be generated, for whatever reason, we shall refer to the usually reliable pseudorandom number generator function labeled RND in most BASIC-language interpreters. The use of pseudorandom numbers rather than a purely random sequence of numbers to simulate the required values is often a matter of much concern. The cyclic nature of pseudorandom number generators is due to the fact that these numbers are generated by deterministic rules. Several methods are available which adequately stretch out these cycles to such an extent that the procedure effectively generates appropriate numbers in as random a manner as possible.

In the case of the TRS-80™ microcomputer, the RND (r) library function will generate pseudorandom numbers in the range $0 \leqslant r \leqslant 1$ when the value passed as the argument (r) is initialized to zero. If you use another BASIC-language interpreter, it would be necessary to read the language manual in order to ensure that the same situation holds true.

Normally, tables of random numbers are accessed when a random number is desired. Since this table would require a certain amount of main memory or external storage (tape or disk) in order to store the table values, the advantages of such a system would be overshadowed by the large storage needs. Since we would appreciate a rapid response to our request for a random number, the use of the RND function will become our sole means of generating such a random number.

More Mathematics

A very important and often cited continuous probability distribution is known as the normal probability distribution. The familiar bell-shaped curve of the nor-

mal probability distribution is one of the most widely seen curves in any text dealing with probability theory. Students are also quite familiar with this curve since
grades are usually considered to follow the shape of the normal probability distribution. However, the importance of the normal probability distribution, for our
purposes, will not be due to its press clippings; rather, it will be used as our standard to examine the concepts we have just discussed, namely, the probability density function, the cumulative distribution function, the mean, and the variance as
each relates to a continuous probability distribution. The practicality of the normal probability distribution lies in the fact that other distributions may be derived
easily or approximated from the normal distribution; for example, the Poisson and
the binomial probability distributions may be approximated from the normal
probability distribution.

Since the normal distribution is placed in the class of probability distributions
known as continuous probability distributions, we may assume that this particular
distribution measures values throughout the range of x; the random variates which
we can generate easily may take on an infinite number of values on the subinterval
of the real number line which encloses the extremes of the distribution. The normal probability distribution has the following basic properties:

1. The curve has but a single peak (maximum value) where $f(x)$ falls off on
   either side of this peak.
2. The curve is symmetrical about the mean or the expected value (μ) of the
   distribution.
3. As the value of x approaches the mean value, the curve slopes upward. As
   the value of x increases from the mean value, the slope of the curve is down-
   ward. In other words, the curve assumes the classical bell shape with the
   center of the bell located at the mean value of the curve.
4. The allowable range of x lies from negative infinity to (positive) infinity,
   that is, the curve is continuous on the range $-\infty \leqslant x \leqslant \infty$.
5. The mean or expected value of the normal probability distribution is denoted
   by the symbol mu (μ).
6. The variance of the normal probability distribution is symbolized by σ^2.
7. The probability density function, $f(x)$, of a random variable X which has a
   normal probability distribution is given by

$$f(x) = \frac{1}{\sigma(2\pi)^{1/2}} \exp^{-(x-\mu)^2/(2\sigma^2)} \tag{2-11}$$

where x may assume any value on the real number line, that is, $-\infty < x < \infty$
and σ is defined as the value of the standard deviation (the root mean square
of the variance).

If the parameters of the normal probability distribution, namely, the mean (μ)
and the standard deviation (σ), take on the values $\mu = 0$ and $\sigma = 1$, then the prob-

ability distribution function is called the standard normal probability distribution. The probability density function, $f(x)$, of the standard normal probability distribution is given by

$$f(z) = \frac{1}{(2\pi)^{1/2}} \exp^{-(z^2/2)} \qquad (2\text{-}12)$$

where the parameter z lies in the range negative infinity to (positive) infinity, that is, the continuum of the real number line.

The transformation from equation 2-11 to equation 2-12 (the standard normal probability distribution) is carried out in the following manner. Initially, the probability density function is assumed to be given as

$$f(x) = \frac{1}{\sigma(2\pi)^{1/2}} \exp^{-(x-\mu)^2/(2\sigma^2)}$$

where $-\infty < x < \infty$ and $\sigma > 0$. Then, the following substitution is applied:

$$z = \frac{x - \mu}{\sigma}$$

This yields the standard normal probability distribution

$$f(z) = \frac{1}{(2\pi)^{1/2}} \exp^{-(z^2/2)}$$

where $-\infty < z < \infty$. We shall refer to equation 2-12 when we require the probability density function, $f(z)$, of a (standard) normal probability distribution.

The corresponding cumulative distribution function, $F(x)$, or, as in the case of the standard normal probability distrubution, $F(z)$, does not exist as a singular function as does the probability density function. Rather, tables are often generated or referred to when particular values for the standard normal probability distribution are required. In order to have an available method from which we may calculate specific values from the cumulative distribution function for the normal probability distribution, we may use the programmed routine given in Figure 2-9.

```
1000  INPUT MU, V, K
1010  T = 0.0
1020  FOR I = 1 TO K
1030    T = T+(RND(0)*(K/2))
1040  NEXT I
1050  X = V*((12/K) ↑ 2)*T+MU
1060  PRINT X
1070  STOP
1080  END
```

FIGURE 2-9: *Programmed routine to obtain values from the C.D.F. for the normal probability distribution*

This programmed routine will generate normally distributed random variates simply by summing k uniformly distributed random variates r_i (where $i = 1, 2, \ldots, k$). The generated r_i's are restricted to the real interval zero to one. Given the mean (μ) and the standard deviation (σ) of our normal probability distribution, we may apply the Central Limit Theorem in order to model mathematically a normal probability distribution. This is accomplished quite easily if we assume that the random variates have an identical probability distribution where:

$$\mu_{r_i} = \theta$$

$$\sigma^2{}_{r_i} = \sigma^2$$

Applying the Central Limit Theorem will yield

$$\lim_{N \to \infty} P \left[a < \frac{\sum\limits_{i=1}^{n} r_i - N\theta}{(N)^{1/2}\sigma} < b \right]$$

or

$$\frac{1}{(2\pi)^{1/2}} \int\limits_{a}^{b} \exp^{-(z^2/2)} dz$$

We now make the following substitutions

$$\mu \left(\sum_{i=1}^{x} r_i \right) = N\theta$$

$$\sigma^2 \left(\sum_{i=1}^{N} r_i \right) = N\sigma^2$$

and

$$z = \frac{\sum\limits_{i=1}^{N} r_i - N\theta}{\sigma(N)^{1/2}}$$

Taking our definition of the standard normal probability distribution, we find that the parameter z becomes a standard normal variate. Applying these equations to our programmed routine variables, we can show that

$$\theta = \frac{a+b}{2} = \frac{0+1}{2} = 0.500$$

$$\sigma = \frac{b-a}{(12)^{1/2}} = \frac{1-0}{(12)^{1/2}} = 0.289 \text{ (approximately)}$$

and finally,

$$z = \frac{\sum\limits_{i=1}^{K} r_i - \frac{K}{2}}{(K/12)^{1/2}}$$

since, by definition

$$z = \frac{x - \mu_x}{\sigma_x}$$

We may now solve for the value x by applying the following:

$$x = \sigma_x \left(\frac{12}{K}\right)^{1/2} \left(\sum\limits_{i=1}^{K} r_i \frac{K}{2}\right) + \mu_x \qquad (2\text{-}13)$$

Equation 2-13 is, therefore, the formula which we shall use in our programmed routine in order to generate normally distributed random variates with a mean equal to μ_x and a variance equal to $\sigma^2{}_x$.

Markov Processes

The concept of Markov processes forms a fundamental property of queueing theory. For this reason, we shall begin our introduction to this process in this chapter and reintroduce it when we study queueing systems in subsequent chapters. The definition and initial investigation of a Markov process was the subject of a paper published by A. A. Markov in 1907. Markov's creation involved a simple form of dependency among random variables which form a stochastic process. The general conceptual matter of a Markov process was eventually wedded into queueing theory and queueing systems as they apply to real-world situations involving waiting lines.

Basically, a Markov process (often referred to as a Markov chain) states that a set of random variables forms a Markov chain if the probability that the value of the next possible random variable depends entirely upon the current random variable and not upon any other previous value for the random variable which preceded this most previous one. Thus, a Markov process forms a chain which effectively links the value of the current random variable to the value of the most recent random variable. In other words, given the value of the most recent random variable, we may chain the entire random sequence merely by working backward one value at a time.

A random sequence of trials $(X_1, X_2, \ldots)$ must satisfy the following properties in order to be considered a stochastic process which forms a Markov chain:

1. Each outcome must be of the finite set of outcomes $(a_1, a_2, \ldots, a_n)$ where this set is called the state space of the system. The state space is defined as the set of possible values or states that a random variable X may assume.
2. The outcome of any trial depends only upon the outcome of the most recent outcome, that is, the outcome preceding this most current one.

The transition probabilities (p_{ij}) of a finite Markov chain may be arranged in an m by n matrix called the transition matrix, that is,

$$
P = \begin{pmatrix}
p_{11} & p_{12} & \cdots & p_{1n} \\
p_{21} & p_{22} & \cdots & p_{2n} \\
 & & \cdot & \\
 & & \cdot & \\
 & & \cdot & \\
p_{n1} & p_{n2} & \cdots & p_{mn}
\end{pmatrix}
$$

If we now take the outcome a_i, there will be a corresponding row $(p_{i1}, p_{i2}, \ldots, p_{in})$ of the appropriate transition matrix P which gives a row vector representing the probability values of all possible values of the next possible outcome. This row vector is actually a probability vector if the components of the vector are nonnegative and their sum is equal to one. Since the components of the vector $(p_{i1}, p_{i2}, \ldots, p_{in})$ are probabilities, their respective values must be, by definition, nonnegative and their sum, again by definition, must be equal to one.

A random variable X relies upon not only its state space and the dependency among its possible values but also another parameter which involves the allowable times at which changes or alterations in the state of the variable may occur. A discrete-time Markov process allows state changes to occur at some finite (or countably infinite) intervals. The analog of the discrete-time Markov chain is the continuous-time Markov process which allows state changes to occur anywhere within a finite (or countable infinite) set of intervals.

The discrete-time Markov chain is a process, $X(t)$, where possible state (value) changes may occur only at the instants of time which are the set of positive integers $(0, 1, \ldots)$; for example, a state change may occur only if $t = 0$ or $t = 6$ and not if $t = 9.1$ or $t = 4.7$. Once a discrete-time Markov process is in a given state, it may remain in that state for a period of time which must be geometrically distributed. The continuous-time Markov process, on the other hand, must be in a given state for a period of time which is exponentially distributed.

Chang[3], Clarke[4], and Kleinrock[5] provide a thorough treatment of the mathematics of the Markov process. Kendall[6] deserves special mention in light of his paper which gave queueing theory the concept of the imbedded Markov chain.

Birth–Death Processes

An important class of Markov chains which has played an equally important role in queueing theory is referred to as the birth-death process. Insofar as queueing theory is concerned, a birth refers to the arrival of a customer into the queueing system, while the term death refers to the eventual departure of a customer from the service facility of a queueing system. The primary consideration, from a mathematical viewpoint, is that state changes (transitions between the possible values of the process) may occur only among states which border each other. In other words, given a set of integers as the state space (a discrete-time Markov process), a birth-death process will manifest itself if the current set is $X_n = i$ and the allowable state transitions (X_{n+1}) are $i - 1, i,$ or $i + 1$ only; given the present state of the system, the only transition which may occur will be to the previous states, the current state, or the next neighboring state.

The particular state of a system at some time t (where $t \geqslant 0$) may be defined in terms of the discrete-time random variable X as $X(t)$. As the time parameter t increases, the resulting probability of $X(t)$ changing becomes the primary goal of the birth-death process. Basically, individual births and deaths occur in a random manner where the average birth-death occurrences must rely only upon the current state of the system. In particular, the following assumptions regarding the birth-death process must be satisfied given $X(t) = x$:

1. Only a single birth or a single death (arrival or departure, respectively) may occur at any given time t.
2. The parameter λ_n (where $n = 0, 1, \ldots$) which defines the average customer arrival rate must be defined in terms of an exponential probability distribution.
3. The parameter μ_n (where $n = 0, 1, \ldots$) which defines the average service rate must be defined in terms of an exponential probability distribution.

[3]W. Chang, "Single Server Queueing Processes in Computing Systems," *IBM Systems Journal* 9(1970): 63–64.
[4]A.B. Clarke and R.L. Disney, *Probability and Random Processes for Engineers and Scientists* (New York: Wiley, 1970), pp. 313–16.
[5]L. Kleinrock, *Queueing Systems, Volume I: Theory* (New York: Wiley, 1975), pp. 26–53, 174–80.
[6]D.G. Kendall, "Stochastic Processes Occurring in the Theory of Queues and Their Analysis by Means of the Imbedded Markov Chain," *Annals of Mathematical Statistics* 24(1953): 338–54.

These assumptions, therefore, define a queuing system with exponentially distributed parameters. Given that there are n customers in the system at time t, then the time interval until the next arrival is also exponentially distributed with an average rate equal to $1/\lambda_n$. This mean value is independent of the time until the next departure. The next departure is, also, exponentially distributed with an average rate equal to $1/\mu_n$. A system described in these terms is called a birth-death process.

Figure 2-10 depicts the rate diagram of the system which we shall examine. The parameters λ_n and μ_n (where $n = 0, 1, \ldots$) are associated with the average birth (arrival rate) and the average death (departure) rate of the system. In particular, our examination of a birth-death process via a rate diagram will center upon the system as it enters and as it leaves a particular state. The rate-in rate-out principle with which we shall be dealing states, in effect, that for any particular state of the system, denoted by n (where $n = 0, 1, \ldots$), the average entering rate must equal the average departing rate. The mathematical equation which symbolizes this principle is called a balance equation (i.e., births are balanced by deaths). If we can form the appropriate balance equations for all possible states of the system, we may then be able to further illustrate how these balance equations relate to the four fundamental queuing system measurements, namely, the average number of customers in the system, the average number of customers in the waiting line, the average time required to complete service, and the average time spent in the waiting line (see chapter 4). Kleinrock[7] and Ross[8] provide the required mathematical background for the birth-death process and its application to queuing theory.

As an example, let us refer to the rate diagram given in Figure 2-10. Initially, we shall assume that the system is in state one (i.e., $n = 1$). The balance equation which we are attempting to form will relate state zero ($n = 0$) to the current state of the system (i.e., $n = 1$). In fact, the only way which the system can enter state n

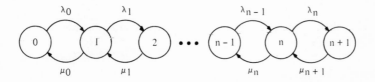

FIGURE 2-10: State-transition rate diagram of a birth-death process

[7]L. Kleinrock, *Queueing Systems, Volume I: Theory* (New York: Wiley, 1975), pp. 53-78.
[8]S.M. Ross, *Introduction to Probability Models* (New York: Academic Press, 1972), pp. 152-57.

= 0 is from its current state ($n = 1$). The average entering rate to state $n = 0$ is equal to μ_1, and since the process cannot enter state $n = 0$ from any of the other given states (except $n = 1$), the average entering rate from any state is equal to zero. The balance equation for state $n = 0$ is given by

$$\mu_1 P_1 = \lambda_0 P_0$$

where P_1 is the steady-state probability of the process being in state $n = 1$, and P_0 is the corresponding steady-state probability that the process is in state $n = 0$. The balance equations for the remaining states are given as:

State	Rate-in	Rate-out
0	$\mu_1 P_1$	$\lambda_0 P_0$
1	$\lambda_0 P_0 + \mu_2 P_2$	$(\lambda_1 \mu_1) P_1$
2	$\lambda_1 P_1 + \mu_3 P_3$	$(\lambda_2 + \mu_2) P_2$
. . .	. . .	. . .
$n - 1$	$\lambda_{n-2} P_{n-2} + \mu_n P_n$	$(\lambda_{n-1} + \mu_{n-1}) P_{n-1}$
n	$\lambda_{n-1} P_{n-1} + \mu_{n+1} P_{n+1}$	$(\lambda_n + \mu_n) P_n$

By manipulating algebraically the above rate balance equations, we may solve for those variables we are most interested in, that is, the probabilities. For example, to solve for the probability of the system being in state $n = 1$, we would manipulate the balance equations

$$\mu_1 P_1 = \lambda_0 P_0$$

in order to obtain the following:

$$P_1 = (\lambda_0 / \mu_1) P_0$$

Since probability theory dictates that the following hold true

$$\sum_{n=0}^{\infty} P_n = 1$$

we may use the balance equation which solves for parameter P_0 in order to state the following:

$$L = \sum_{n=0}^{\infty} n P_n \tag{2-14}$$

$$L_q = \sum_{n=s}^{\infty} (n - s) P_n \tag{2-15}$$

$$W = L / \lambda \tag{2-16}$$

$$W_q = L_q / \lambda \tag{2-17}$$

where parameter λ is equal to

$$\sum_{n=0}^{\infty} \lambda_n P_n$$

The results in equations 2-14 through 2-17 will be used later during our study of the various measurements of queueing system perfomrance. For now, we may state the following definitions for the parameters solved for in these equations:

L = average number of customers in the system
L_q = average number of customers in the waiting line
W = average time required to complete service
W_q = average time spent in the waiting line

Before we close this chapter, it would be useful to review the concepts we have examined by solving a few problems.

Example Problems

Problem 1

Our first problem, which deals with the elements of probability theory previously examined, involves the tossing of the fair six-sided die. The given sample space of this particular experiment is denoted by

$$S = (1, 2, 3, 4, 5, 6)$$

where each of the possible outcomes has a probability of occurrence equal to 1/6. Our problem will be to find the distribution, the expected value, the variance, and the standard deviation of the random variable X (where X will refer to twice the value of the face-up value of the die). The parameters of the random variable X, therefore, can be denoted by

$$\begin{aligned}
X_1 &= 2 \\
X_2 &= 4 \\
X_3 &= 6 \\
X_4 &= 8 \\
X_5 &= 10 \\
X_6 &= 12
\end{aligned}$$

that is, $X(S) = (2, 4, 6, 8, 10, 12)$ and each of the values has a probability of occurrence equal to 1/6. Thus, the distribution of the random variable X is given by:

x_i	2	4	6	8	10	12
$f(x_i)$	1/6	1/6	1/6	1/6	1/6	1/6

The calculations which solve for the expected value or mean as well as the variance and the standard deviation are below. The expected value of the distribution is given by:

$$\mu = \sum_{i=1}^{n} x_i f(x_i)$$
$$\mu = 2(1/6) + 4(1/6) + 6(1/6) + 8(1/6) + 10(1/6) + 12(1/6)$$
$$\mu = 7$$

The variance is calculated by applying the following:

$$\sigma^2 = \sum_{i=1}^{n} x_i^2 f(x_i) - \mu^2$$
$$\sigma^2 = (4(1/6) + 16(1/6) + 36(1/6) + 64(1/6) + 100(1/6) + 144(1/6)) - (7)^2$$
$$\sigma^2 = 11.7 \text{ (approximately)}$$

The standard deviation is calculated by taking the root mean square of the variance, or

$$\sigma = (\sigma^2)^{1/2}$$
$$\sigma = (11.7)^{1/2}$$
$$\sigma = 3.4 \text{ (approximately)}$$

Although the previous example problem seems rather simple, the concern, at this time, is that we are able to calculate these statistical properties. Subsequent problems in this section will use the fundamental concepts of these statistical properties to show the relationships of various notions of probability theory.

Problem 2

Our second example problem involves a more complex probability function as well as more complex calculations. Assume that the random variable X is a continuous random variable which has the following distribution:

$$f(x) = \begin{cases} (1/6)x + k & \text{for } 0 \leqslant x \leqslant 3 \\ \\ 0 & \text{elsewhere on the real numbers} \end{cases}$$

First we must evaluate parameter k, and then determine the probability P that the random variable X lies in the real interval ($1 \leqslant X \leqslant 2$).

Since the function described is a continuous probability function, the cross-hatched area of Figure 2-11 will have an area equal to one. The area forms a trapezoid with parallel lines (bases) which have a length equal to parameter k and $k + (1/2)$. The height of the trapezoid is equal to 3. Therefore, the area is equal to

$$A = 1 = 1/2(k + k + 1/2)3$$

and the parameter k is equal to $1/12$.

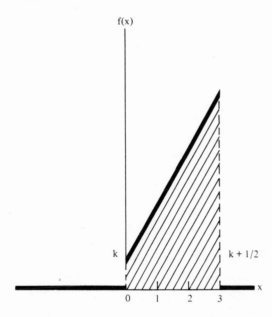

f(x)

k

k + 1/2

0   1   2   3

x

FIGURE 2-11: Graph of the function f (x) = (1/6) x + k for 0 ⩽ x ⩽ 3

The probability $P(1 \leqslant X \leqslant 2)$ will be equal to the cross-hatched area of the graph depicted in Figure 2-12. Since $f(1)$ and $f(2)$ are given as

$$f(1) = 1/6 + 1/12 = 3/12$$
$$f(2) = 1/8 + 1/12 = 5/12$$

then $P (1 \leqslant X \leqslant 2)$ will be equal to $1/2 (3/12 + 5/12)$ or $1/3$.

Problem 3

This last problem involves the random variable X which has the following probability distribution:

$$f(x) = \begin{cases} k & \text{for } a \leqslant x \leqslant b \\ \\ 0 & \text{elsewhere on the real line} \end{cases}$$

This random variable is, in fact, a continuous random variable which is uniformly distributed on the interval I as shown in Figure 2-13. The cross-hatched area is equal to one, therefore,

$$k(b - a) = 1 \text{ or } k = \frac{1}{b - a}$$

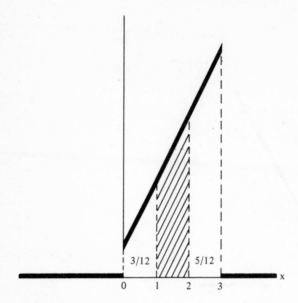

FIGURE 2-12: Graph of P (1 ≤ x ≤ 2)

The expected value or mean of this particular probability distribution is given as

$$\mu = \frac{a + b}{2}$$

or the point on the interval I which is halfway between a and b. For example, if $a = 2$ and $b = 3$, then the mean would be equal to:

$$\mu = \frac{2 + 3}{2} = 2.5$$

The cumulative distribution function is defined by:

$$F(k) = P(X \leqslant k)$$

Therefore, $F(k)$ relates to the area under the graph of the probability density function (f) which lies to the left of $x = k$. Since the random variable X is defined as being informly distributed on the interval I, where

$$I = (a \leqslant x \leqslant b)$$

the graph of the cumulative distribution function is given as shown in Figure 2-14.

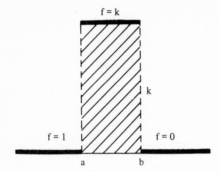

FIGURE 2-13: Probability density function of the uniformly distributed random variable $f(x) = k$ *for* $a \leqslant x \leqslant b$

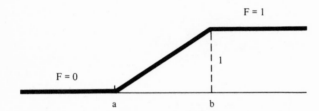

FIGURE 2-14: Graph of the cumulative distribution function $F(x) = P (x \leqslant k)$

3

Probability Distributions

Of the many possible probability distributions available, we must devote attention to only a few which can later be implemented in terms of their application to queueing systems. This chapter introduces the probability distributions which we shall use later in our study of the various types of real-world queueing systems. They are as follows:

1. Uniform distribution
2. Binomial distribution
3. Poisson distribution
4. Exponential distribution
5. Gamma distribution
6. Erlang distribution

Each of the above probability distributions will be considered in terms of the concepts of probability theory previously examined. First, a particular probability distribution will be illustrated in terms of its probability density function and its corresponding cumulative distribution function; second, the necessary generation of the required random variates for the particular probability distribution will be presented; and third, the appropriate formulas which will calculate the mean and the variance of the probability distribution under study will be given. With regard to generation of the random variates of a particular probability distribution, an appropriate programmed routine will also be presented which will effectively "computerize" this task. The values for the corresponding mean and variance of the given distribution will also be achieved via a programmed routine.

The programmed routines presented in this chapter will be used again in chapters 5 and 6 when we apply a particular probability distribution to a real-world queueing system. At the appropriate time, these routines will be repeated.

43

Uniform Probability Distribution

In order to introduce the concepts of probability theory as they relate to probability distributions in as painless a manner as possible, we shall first consider the uniform or rectangular probability distribution. This distribution is often used when the random occurrence of events which are described on the real number line are needed; in other words, if a random variable X can be fitted to a uniform probability distribution, then its values are restricted to a particular subinterval of the real number line. Our first requirement, therefore, is to present the probability density function and the corresponding cumulative distribution function of the uniform probability distribution. From these two functions, we shall introduce the generation of the random variates of this probability distribution as well as the mean and the variance of the uniform probability distribution.

The probability density function, $f(x)$, for the uniformly distributed random variable X is given as

$$f(x) = \frac{1}{b - a} \tag{3-1}$$

where a and b are represented as an ordered pair of real numbers, that is, (a, b). Furthermore, parameter a must be less than parameter b and $a < x < b$. Graphi-

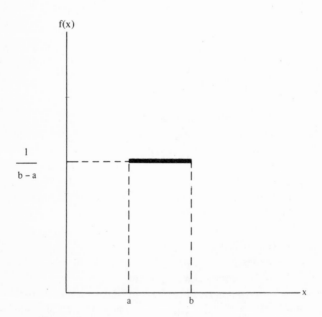

FIGURE 3-1: Probability density function of the uniform probability distribution

cally, the probability density function of the uniformly distributed random variable X (equation 3-1) is illustrated in Figure 3-1.

The corresponding cumulative distribution function, $F(x)$, for the uniformly distributed random variable X is found by integrating the given probability density function (equation 3-1) over the interval (a, b) which yields

$$F(x) = \frac{x - a}{b - a} \qquad (3-2)$$

on the ordered pair of real numbers (a, b) where parameter a is less than parameter b, as was the case in the previous definition for the probability density function. Figure 3-2 illustrates the corresponding graphical representation of the cumulative distribution function of the uniformly distributed random variable X.

The equations which define the probability density function and the corresponding cumulative distribution function of any particular probability distribution are, of course, the necessary first steps in further descriptions of the remaining statistical properties of the distribution under study. Graphical illustrations of these functions are also a convenient tool as is a programmed routine which will calculate the coordinate points which correspond to the graph. Figure 3-3 is the programmed routine which will allow us to receive these coordinate points merely by entering the values for the given ordered pair of real numbers (a, b).

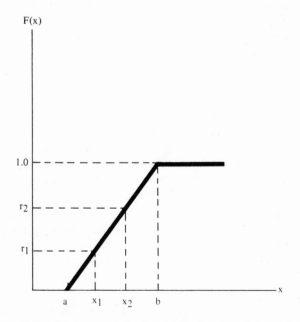

FIGURE 3-2: Cumulative distribution function of the uniform probability distribution

```
1000  INPUT"ENTER PARAMETER A AND PARAMETER B AS A, B"; A, B
1010  IF A > B THEN 1050
1020    PRINT"PARAMETER A MUST BE GREATER THAN PARAMETER B"
1030    PRINT
1040    GOTO 1000
1050  PRINT"P.D.F        C.D.F"
1060  FOR X = 1 TO 10
1070    PX = 1.0 / (B-A)
1080    CX = (X-A) / (B-A)
1090    PRINT PX, CX
1100  NEXT X
1110  GOTO 1000
1120  END
```

FIGURE 3-3: Programmed routine to generate values for the uniform probability density and cumulative distribution functions

Our next step is to define a method by which the random variates which are based on our given probability distribution will be generated. There are a number of convenient techniques we can use. Initially, we shall use the inverse transformation method (see chapter 2), since it offers a simple mathematical approach. In order to be able to generate random variates for a given probability distribution, we must first have available that distribution's probability density function. From this we form the corresponding cumulative distribution function and then generate the random variates for the given probability distribution. In the case of the uniform probability distribution, the probability density function is given in equation 3-1, while the corresponding cumulative distribution function is given in equation 3-2.

Since $F(x)$, the cumulative distribution function, is forced to cover the range zero to one of the real number line, we may generate uniformly distributed random numbers in the range zero to one simply by calling the RND library function of our BASIC-language interpreter. Although the RND function will give pseudorandom numbers as output, this will not present any problems in defining the inverse transformation method of generating random variates.

Since $0 \leqslant F(x) \leqslant 1$ and the generated random number (r) has a range of zero to one, we may set $F(x) = r$. Because parameter x is uniquely defined by $F(x) = r$, it follows that for any value of r which can be generated, it is possible to find a corresponding value of x by the inverse transformation of the function F through

$$x = F^{-1}(r)$$

where $F^{-1}(r)$ is a mapping of the random number r onto the unit interval into the domain of the parameter x.

Returning to our uniform probability distribution, we can obtain, from the probability density function of equation 3-1, the corresponding cumulative distribution function given in equation 3-2. Taking the inverse transformation of equation 3-2 yields the following:

$$f(x) = \frac{1}{b - a} \quad \text{where } a < x < b$$

$$F(x) = \frac{x - a}{b - a} \quad \text{where } 0 \leqslant F(x) \leqslant 1$$

And, by the inverse transformation method, as follows:

$$x = a + (b - a)r \quad \text{where } 0 \leqslant r \leqslant 1 \tag{3-3}$$

We are now in a position, mathematically, to generate a set of random numbers r which will correspond to the defined range of $F(x)$, namely, zero to one. Figure 3-2 illustrates the fact that for a particular set of generated random numbers over the cumulative distribution function, a unique uniformly distributed random variate is given over the interval (a, b).

Given the above information, it is now a simple matter to write a short programmed routine which will generate the random variates of a uniformly distributed random variable. Figure 3-3 accomplishes this task.

Two important statistical quantities which relate much more information about a particular probability distribution than do the previously discussed functions $f(x)$ and $F(x)$ are the mean or the expected value of the probability distribution and the variance of the distribution. The attendant derivations of these statistical properties are covered in much of the literature. Our primary interest lies in the fact that formulas for these properties do exist.

The mean or expected value of a uniformly distributed random variable X is given by

$$\mu = \frac{b + a}{2} \tag{3-4}$$

while the variance of a uniformly distributed random variable X is given by:

$$\sigma^2 = \frac{(b - a)^2}{12} \tag{3-5}$$

Under normal conditions, the mean and the variance of a particular probability distribution are the known or given parameters. The actual values for the ordered pair of real numbers (a, b), for the uniform probability distribution, are, quite often, unknown. Therefore, in a situation where the values (a, b) must be known, given the values for the mean (μ) and the variance (σ^2), some method must be available which will allow us to calculate the unknown values.

For the uniformly distributed random variable X, defined over the interval $(a,$ $b)$, and given the values for its mean and variance, the following restructured formulas exist:

$$a = \mu - (3\sigma^2)^{1/2}$$

$$b = 2\mu - a$$

Binomial Probability Distribution

Although the binomial probability distribution is not used in queueing theory as a representative probability distribution, its description is, nevertheless, important in order to serve as an introduction for the next probability distribution which we shall examine, namely, the Poisson probability distribution. However, our next order of business in this chapter is with the binomial probability distribution and its statistical properties.

If we were to perform an experiment which had but two possible outcomes, we would be performing what is referred to as a Bernoulli trial. These specially named experiments were among the first probabilistic activities carried out by probability theorists. Basically, a Bernoulli trial is an independent experiment where the outcomes of the experiment are labeled as either success or failure (win or lose).

Let us further discribe our Bernoulli trial by calling the probability of a success p (where $0 \leqslant p \leqslant 1$) and the probability of a failure q (where, of course, $p + q = 1$). In addition, we shall assume that if the probability of success or failure is constant, given a sequence of identically conducted trials, then we may call this sequence of trials a Bernoulli sequence. A Bernoulli random variable X is, therefore, a random variable which may assume only one of two possible states or values: a value of one (for a success) or a value of zero (for a failure).

If we were to conduct a Bernoulli trial exactly n times, the random variable X which counts the number of successes would be called a binomial variable with parameters n (for the number of trials) and p (for the probability of success). This random variable may assume only the discrete values $0, 1, \ldots, n$. From this description, we can state the probability density function, $f(x)$, of a binomial random variable X as

$$f(x) = \binom{n}{x} p^x q^{n-x} \tag{3-6}$$

where $\binom{n}{x}$ is defined mathematically as

$$\binom{n}{x} = \frac{n(n-1)(n-2)\ldots(n-x+1)}{x!}$$

where the parameters x and n must be positive integers and $x \leqslant n$. These numbers are referred to as the binomial coefficients according to the Binomial Theorem below:

$$(p+q)^n = \sum_{x=0}^{n} \binom{n}{x} p^x q^{n-x} \qquad (3\text{-}9)$$

The expansion $(p+q)^n$ has the following properties:

1. There will be $n+1$ terms.
2. The sum of the exponents of p and q in each term will equal the value of n.
3. The exponents of parameter q will decrease term by term from n to zero.
4. The exponents of parameter p will increase term by term from zero to n.
5. The coefficients of any term are $\binom{n}{k}$ where parameter k is the exponent of either p or q.

The coefficients of the successive powers of $p+q$ can be arranged into a triangular array (matrix) of numbers known as Pascal's triangle, that is:

$$(p+q)^0 = 1$$
$$(p+q)^1 = p+q$$
$$(p+q)^2 = p^2 + 2pq + q^2$$
$$(p+q)^3 = p^3 + 3p^2q + 3pq^2 + q^3$$
$$(p+q)^4 = \text{and so on}$$

Pascal's triangle has many interesting properties in its own right, however it is not in our best interest, at this time, to consider these characteristics.

As an example of an application of a binomial probability distribution, we shall return to our earlier die-tossing experiment. If we were to toss a fair six-sided die 10 times and call a toss a success if either a one or a six appeared as the face-up value, then the probability of a success (p) would be given as

$$p(1, 6) = 1/3$$

and the corresponding probability of a failure (q) would be given as

$$q(2, 3, 4, 5) = 2/3$$

that is,

$$q = p - 1 = 0.333 \ldots -1 = 0.666 \ldots$$

The probability that a one or a six would appear exactly three times in a 10-toss trial would be given as

$$f(x;n;p) = f(3;10;1/3) = \binom{10}{3}(1/3)^3(2/3)^4$$

or 0.2561 (approximately).

The probability that neither a one nor a six would appear as the face-up value would be:

$$q^7 = (2/3)^7 = 0.0585 \text{ (approximately)}$$

Hence, the probability that a one or a six occurs at least one time is given by:

$$1 - q^7 = 0.9415 \text{ (approximately)}$$

The calculations of other problems involving die tossing or experiments which have outcomes only considered as successes or failures could go on endlessly. Later we shall have at our disposal a programmed routine which will handle the various mathematical manipulations of the binomial probability distribution. For now, however, we have a pressing need to look at some of the more important statistical properties of the binomial probability distribution.

The mean, the variance, and the standard deviation values for the binomial probability distribution are given as:

$$\text{mean} = \mu = np \tag{3-8}$$

$$\text{variance} = \sigma^2 = npq \tag{3-9}$$

$$\text{standard deviation} = \sigma = (npq)^{1/2} \tag{3-10}$$

It should be noted from the above that the variance (σ^2) is always less than the mean value (μ). The parameters used to calculate these properties are the number of trials (n), the probability of success (p), and the probability of failure (q).

Usually, the respective values for n, p, and q are not the known parameters of the distribution. It is more often the case that the mean value and the variance are the given or the known parameters. This is true not only for the binomial probability distribution, but for most of the other probability distributions which we shall encounter later in this chapter. Therefore, the formulas for the mean and the variance must be rearranged in such a manner as to accommodate these known parameters. For the binomial probability distribution, the following rearrangements hold:

$$p = \frac{\mu - \sigma^2}{\mu}$$

$$n = \frac{\mu^2}{\mu - \sigma^2}$$

In order to calculate the mean or expected value which corresponds to any one particular face-up value appearing in our die-tossing experiment, we would assume n trials having a probability of success equal to p. In other words, if we take 100 trials, where the probability of success of any one value appearing face-up is equal to $1/6$, then the mean or expected value would be equal to:

$$\mu = np = 100(1/6) = 16.666 \ldots$$

The value attained would be the theoretical expected value of any face-up value appearing by tossing a fair six-sided die.

The variance of any particular face-up value appearing after 100 tosses or trials would be:

$$\sigma^2 = npq = 100(1/6)(5/6) = 13.9 \text{ (approximately)}$$

Binomial random variates may be generated by the application of the rejection method. Initially, known values for parameters n and p must be given or calculated. The process which effectivley generates the required random variates consists of the generating of n pseudorandom numbers after the first (x_0) random variate is initialized to zero. For each allowable pseudorandom number r_i (where $i = 1, 2, \ldots, n$), the x_i variable in the programmed routine is incremented as follows:

$$x_i = x_{i-1} + 1 \qquad \text{if } r_i \leq p$$

$$x_i = x_{i-1} \qquad \text{if } r_i > p$$

that is, if the ith pseudorandom number is less than the probability of success (p), then the x_i generated random variate is equal to the previously generated random variate plus one. On the other hand, if the generated pseudorandom number is greater than the probability of success, then the value assigned to x_i will be equal to the previous x_i value, that is, no change will take place.

After n such numbers have been generated, the x_n value will be equal to the binomial variate x. Figure 3-4 is the programmed routine which accomplishes this generation of binomially distributed random variates. Figure 3-5 shows the sample output from this programmed routine.

```
1000  INPUT"N,P";N, P
1010  PRINT"SUCCESSES      F(X)              CUMULATIVE"
1020  X = 1.0 - P
1030  BD = X ↑ N
1040  C = BD
1050  FOR K = 1 TO N+1
1060     IF BD < 10E-6 THEN 1100
1070       PRINT K-1, BD, C
1080       IF ABS (C-1.0) < 10E-10 THEN 1140
1090         IF K > N THEN 1130
1100           BD = (N-K+1)/K * BD
1110           BD = BD * P/X
1120           C = C + BD
1130  NEXT K
1140  GOTO 1000
1150  END
```

FIGURE 3-4: Programmed routine to generate binomial random variates

N,P? 12,.75

SUCCESS	F(X)	CUMULATIVE
2	.0000	.0000
3	.0004	.0004
4	.0024	.0028
5	.0115	.0143
6	.0405	.0548
7	.1032	.1580
8	.1936	.3516
9	.2581	.6097
10	.2323	.8420
11	.1267	.9687
12	.0317	1.0000

N,P? 32,.3333

SUCCESS	F(X)	CUMULATIVE
0	.0000	.0000
1	.0000	.0000
2	.0003	.0003
3	.0014	.0017
4	.0052	.0069
5	.0146	.0215
6	.0328	.0543
7	.0609	.1152
8	.0952	.2104
9	.1270	.3374
10	.1460	.4834
11	.1460	.6294
12	.1278	.7572
13	.0983	.8565
14	.0667	.9232
15	.0400	.9532
16	.0213	.9745
17	.0100	.9845
18	.0042	.9887
19	.0015	.9902
20	.0005	.9907

FIGURE 3-5: Sample output from programmed routine of Figure 3-4

Poisson Probability Distribution

An important probability distribution which enjoys a singularly important role in many real-world and naturally occurring phenomena is the Poisson process. Named in honor of Simon Denis Poisson, a French mathematician known primarily for his work involving probability theory, a Poisson process may, in fact, be the vehicle which can be used as an effective mathematical tool in such diverse fields as biology, meteorology, radiology, and the rapidly expanding field of telecommunications. Queueing theory also makes use of the Poisson probability distribution with regard to the modeling of customer arrival patterns. From a purely mathematical position, the Poisson probability distribution may be effectively utilized as a method to approximate the binomial probability distribution. Before we leap ahead to a description of a Poisson process and its relationship to queueing theory, however, it would be worthwhile to illustrate the role of this particular probability distribution with regard to the binomial probability distribution as well as its role as an interesting and important mathematical tool for which many real-world events follow.

In general terms, the Poisson process describes the occurrence of certain random events during a particular interval of time (t) over a given time period (T). In particular, the occurrence of these events over time is considered to be of a random fashion. The key words in this initial description are events and random. Our previous definitions for these two words will fit quite well into our examination of the activity of a Poisson process.

Let us take a real-world example of what may be considered a Poisson process. The number of aircraft arriving at a busy airport may, during a very long period of time T, be rather staggering. However, if we view only a small portion of time (t) during any particular interval of our primary period of time (T), we would see that the probability of having a large number of aircraft arriving at the airport would be rather small indeed. In other words, the probability of having simultaneous arrivals of two or more aircraft in one second of time during an observation period of one hour of time would be very small.

From a probabilistic viewpoint, simultaneous events are, in the real world, often considered to be negligible. Real-world events, especially those which can be modeled mathematically by a Poisson process, usually have events occurring one after another with some definable period of time separating each event.

This real-world phenomenon may be rather difficult to envision, since we usually find ourselves in the midst of others like us who demand immediate service upon arrival at some servicing facility. We may well ask ourselves, how many times have we come to a revolving door at the same instant of time that someone else arrived there? We must, however, keep in mind that these aberrations of probability theory are quite possible (albiet, quite small when we consider only their overall probability of occurrence). In addition, we must take into account a time interval which describes a small enough period of time in order for the real-world process to become a mathematical model of a Poisson process.

Since a Poisson process is, in fact, describable in terms of a probability distribution, we now turn our attention to the mathematical aspects of this particular
process. Mathematically, the Poisson process may be defined in terms of the probability of a random variable X which takes on the particular value x, that is,

$$P(X = x) = p_x(\lambda)$$

where the lambda (λ) is called the parameter of the distribution. The Poisson probability distribution with parameter λ can be defined by the equation

$$P(x;\lambda) = \frac{\lambda^x \exp^{-x}}{x!} \qquad (3\text{-}11)$$

where $x = 0, 1, \ldots$ and $\lambda > 0$ and λ is also a constant. Of course, $\exp = 2.71828\ldots.$

Since we are primarily interested in the probability of having exactly x events
occurring in some interval of time (t), the Poisson probability distribution will be
defined in terms of this interval of time as

$$P_x(t) = \frac{\lambda t^x \exp^{-\lambda t}}{x!}$$

where $P_x(t)$ will be called the probability of exactly x events occurring during the
given interval of time t. Parameter λ must be a positive non-zero constant which
signifies the average number of events which will be allowed to occur during our
observation interval of time t.

Figure 3-6 graphs typical Poisson process values for parameter λ. The coordinate points for these graphs were produced by the programmed routine in Figure
3-7.

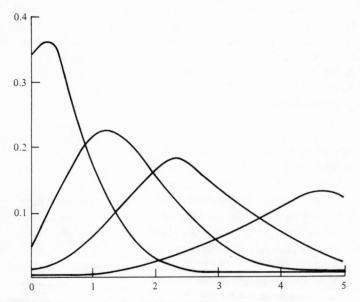

FIGURE 3-6: Graph of typical Poisson process values

```
1000  INPUT"LAMBDA"; LM
1010  PRINT" N          P"
1020  FOR KV = 0 TO 10
1030    KF = 1
1040    FOR KK = 1 TO KV
1050     KF = KF * KK
1060    NEXT KK
1070    P = ((LM ↑ KV) / KF) * (EXP (-1 * LM))
1080    PRINT KV, P
1090  NEXT KV
1100  GOTO 1000
1110  END
```

LAMBDA? 1

N	P
0	.3679
1	.3679
2	.1839
3	.0613
4	.0153
5	.0031
6	.0005
7	.0000
8	.0000
9	.0000
10	.0000

LAMBDA? 3

N	P
0	.0498
1	.1494
2	.2240
3	.2240
4	.1680
5	.1008
6	.0504
7	.0216
8	.0081
9	.0027
10	.0000

FIGURE 3-7: Programmed routine to generate Poisson values

LAMBDA? 5

N	P
0	.0067
1	.0337
2	.0842
3	.1404
4	.1755
5	.1755
6	.1462
7	.1044
8	.0653
9	.0363
10	.0181

LAMBDA? 10

N	P
0	.0000
1	.0000
2	.0023
3	.0076
4	.0189
5	.0378
6	.0631
7	.0901
8	.1126
9	.1251
10	.1251

FIGURE 3-7: (Continued)

At this point, we should compare the graph of the Poisson probability distribution with that of the previously described uniform probability distribution.
The primary difference, from a graphical viewpoint, between these two processes
involves the probability of an occurrence of a given event. The uniform probability
distribution allows particular events to occur with the same probability regardless
of the time period under consideration. The Poisson probability distribution, on
the other hand, allows a number of events (n) to occur during a given time period;
in addition, these particular events are allowed to occur in a random manner rather
than in a known manner, as is the case for the uniform probability distribution.
A comparison of the uniform probability distribution graph (Figure 3-1) and the
Poisson probability distribution graph (Figure 3-6) clearly demonstrates their dif-
ferences.

The Poisson probability distribution is known as a discrete probability distribution which counts values throughout the given range of x. Additionally, the Poisson probability distribution has a positive probability that the random variable X may assume a value of zero. Contrast this fact with the normal probability distribution where, theoretically, $x = 0$ does not exist, that is, the normal probability distribution does not have a positive probability that the random variable X may assume the value $x = 0$.

A real-world example of this condition follows. If parameter x counts the number of patrons arriving at a movie theater ticket booth every 15 seconds, then according to the definition of a Poisson process, there exists the positive probability that no patrons will arrive at the ticket booth in the given period of time t. Thus, arrivals to the ticket booth during a prescribed period of time t may vary from zero to infinity. The arrival of an infinite number of patrons is, of course, taken in the probabilistic sense rather than in the real-world sense. The population which supplies the patrons is not infinite; however, in order to handle some of the mathematics of probability theory properly and easily, it is best to assume an infinite source population. We shall deal more with this particular fact under fundamental aspects of queueing systems in chapter 5.

Figure 3-6 illustrated two basic properties of Poisson probability distribution:

1. A single peak or maximum value
2. No symmetry about its mean value, in fact, there are more values of x greater than the mean value

This skewing of the probability distribution curve about the mean value was not evident in the normal probability distribution. In fact, the curve of the normal probability distribution assumed its maximum value at the mean value of the curve. The curve also intersects the y-axis, that is, $f(x)$, at a value which is numerically equal to $\exp^{-x}$; in other words, the random variable X will assume a value of $x = 0$ at the point $y = \exp^{-x}$.

For a Poisson process to be the applicable probability distribution which will effectively model the random occurrence of particular events during a given period of time t, it is necessary that the probability of x such events in a time period of length t adhere to the following restrictions:

1. The probability of more than one event occurring during a short period of time t must essentially be zero, that is, simultaneous events are considered to be negligible.
2. The probability of one event occurring during a short period of time t is proportional to the length of the time period T under current observation.
3. The number of events occurring during any two different time periods must be independent.

These properties will, therefore, guarantee that equation 3-11 holds true, that is, the Poisson process may be defined as

$$P_x(t) = \frac{\lambda t^x \exp^{-\lambda t}}{x!}$$

where $x = 0, 1, \ldots$ and $\lambda > 0$ and λ is a constant.

We mentioned earlier that the Poisson probability distribution is often utilized as an approximation to the binomial probability distribution. From this fact, supported by the output of Figure 3-8, we can further define some of the important statistical properties of the Poisson probability distribution. These properties include the mean or the expected value of the distribution and the variance of the distribution. The output in Figure 3-9 was obtained from the programmed routine of Figure 3-8.

```
1000  INPUT"N,P"; N,P
1010  PRINT"K            BINOMIAL            POISSON"
1020  X = 1.0 - P
1030  BD = X ↑ N
1040  NP = N * P
1050  PD = EXP (-NP)
1060  FOR K = 1 TO 5
1070    PRINT K-1, BD, PD
1080    IF K > N THEN 1120
1090      BD = (N-K + 1.0)/K * BD
1100      BD = BD * (P/Q)
1110      GOTO 1140
1120      BD = 0.0
1130      PD = PD * NP/K
1140  PD = PD * NP/K
1150  NEXT K
1160  GOTO 1000
1170  END
```

FIGURE 3-8: Relationship of binomial to Poisson distributions

N,P? 100,.01

K	BINOMIAL	POISSON
0	.3660	.3679
1	.3697	.3679
2	.1848	.1839
3	.0610	.0613
4	.0149	.0153
5	.0029	.0031

FIGURE 3-9: Sample output from Figure 3-8

N,P? 80,.025

K	BINOMIAL	POISSON
0	.1319	.1353
1	.2706	.2707
2	.2741	.2707
3	.1827	.1804
4	.0902	.0902
5	.0352	.0360

N,P? 100,.05

K	BINOMIAL	POISSON
0	.0059	.0067
1	.0312	.0337
2	.0812	.0842
3	.1396	.1404
4	.1781	.1755
5	.1800	.1755

FIGURE 3-9: (Continued)

If we were to take the binomial probability function

$$f(x;n;p) = \binom{n}{x} p^x (1-p)^{n-x}$$

where $x = 0, 1, \ldots, n$ for a fixed value of parameter n, and force the following restrictions

$$n \to \infty$$

$$p \to 0$$

$$\lambda = np \text{ and is a constant}$$

then the equation for the binomial probability distribution would become the equation for the Poisson probability distribution, that is:

$$f(x) = \frac{\lambda^x \exp^{-x}}{x!}$$

The attendant mathematics for this derivation are treated quite thoroughly in the texts listed in the Bibliography under Ross, Clarke and Disney. It will be sufficient here to say that if a binomial random variable X which has the parameters n and p where parameter n approaches infinity as parameter p approaches zero (where the product np remains constant), then the approximation

$$P(X = x) = \exp^{-np} \frac{(np)^x}{x!}$$

where $x = 0, 1, \ldots, n$, holds true.

Given that $\lambda = np$ is some constant (where the product np was defined earlier as the expected value of the binomial probability distribution), we are now in a position to describe additional statistical properties, namely, the mean or expected value and the variance of the Poisson probability distribution.

The mean or expected value is taken from the corresponding product (np) for the binomial probability distribution where the product (np) is a constant as parameter n approaches infinity as parameter p approaches zero, that is:

$$\text{mean} = \mu = \lambda = np \tag{3-12}$$

Likewise, the variance is given by

$$\text{variance} = \sigma^2 = \lambda = np \tag{3-13}$$

again, as parameter n approaches infinity as parameter p approaches zero while the product np remains constant.

An additional statistical property often required in probability work is that of the standard deviation. It is given as

$$\text{standard deviation} = \sigma = (\lambda)^{1/2} \tag{3-14}$$

where parameter λ is equal to the product of the parameters (np) from the binomial probability distribution.

Rather than present the often rigorous proofs for these statistical quantities, we shall use a programmed routine which will, for various values of parameters n and p, show the corresponding relationship between the Poisson probability distribution and the binomial probability distribution. Figure 3-9 accomplishes this task.

Exponential Probability Distribution

Just as the Poisson probability distribution is widely used as an approximation to many real-world situations (especially with regard to queueing systems), so too is the exponential probability distribution used as a good approximation to real-world mathematical models in queueing systems. In addition to being a good fit to queueing models, this distribution is rather easy to work with in terms of its mathematics.

The exponential probability distribution is used when measurements along a continuum are required and is therefore classified as a continuous probability distribution. The normal probability distribution was another continuous probability distribution, while the Poisson probability distribution was classified as a discrete probability distribution. In addition, the exponential probability distribution is based in independent occurrences of some particular events, that is, it too is classified as a random process which follows the rules we set forth in our previous discussion concerning the laws and concepts of probability theory.

In particular, the exponential probability distribution may be expressed by the probability density function

$$f(x) = a\exp^{-ax} \quad \text{for } x \geqslant 0 \tag{3-15}$$

where parameter a is a constant greater than zero and is independent of any other factors which may tend to influence the distribution.

As a continuous probability distribution, the exponential distribution measures values throughout the range of parameter x . The Poisson probability distribution; had a positive probability that parameter x would assume a value of zero. So too does this particular property exist with the exponential probability distribution; in fact, the value which corresponds to the probability of x assuming a value of zero is equal to the value of parameter a. As the value of parameter x decreases, the probability density function, $f(x)$, increases in value, attaining its greatest value when $x = 0$. The programmed routine in Figure 3-10 will output the necessary points to construct the curve for the exponential distributions.

```
1000  INPUT"LAMBDA";L
1010  INPUT"MAX T";T
1020  PRINT"TIME          F(T)"
1030  FOR TV = 0 TO T STEP 0.5
1040    FT = 1.0-EXP (-1 * (L * TV))
1050    PRINT TV, FT
1060  NEXT TV
1070  GOTO 1000
1080  END
```

LAMBDA? 1
MAX T? 6

TIME	F(T)
0.0	.000
0.5	.394
1.0	.632
1.5	.777
2.0	.865
2.5	.918
3.0	.950
3.5	.970
4.0	.982
4.5	.989
5.0	.983
5.5	.996
6.0	.998

FIGURE 3-10: Programmed routine to generate exponential variates

LAMBDA? 2
MAX T ? 6

TIME	F(T)
0.0	.000
0.5	.632
1.0	.865
1.5	.950
2.0	.982
2.5	.993
3.0	.998
3.5	.999
4.0	.999
4.5	.999
5.0	.999
5.5	.999
6.0	.999

LAMBDA? 3
MAX T ? 6

TIME	F(T)
0.0	.000
0.5	.777
1.0	.950
1.5	.989
2.0	.998
2.5	.999
3.0	.999
3.5	.999
4.0	.999
4.5	.999
5.0	.999
5.5	.999
6.0	.999

FIGURE 3-10: (Continued)

The exponential probability distribution has found wide use, since it is able to model mathematically many real-world situations regarding queueing systems. For example, the arrival of airplanes at an airport and the arrival of patients into a hospital admitting room can be fitted to the exponential probability distribution. The mathematical definition of this type of arrival behavior can be stated as follows: If the probability of an event occurring in a particularly small interval of time t is itself very small (for example, births, deaths, and accidents), and if the occurrence of this particular event is independent of the occurrence of other events, then it can be said that the time interval between the occurrence of any two successive events is exponentially distributed. In fact, the probability that one event will occur during some interval of time t is equal to $a\Delta t$. The probability that more than one event will occur during some interval of time t approaches zero (that is, becomes negligible) as Δt tends to zero.

The assumption that one particular event is independent of the occurrence of other events involves the memoryless property of a given probability distribution, that is, the process effectively "forgets" its history or what has occurred prior to this particular instant in time. The memoryless property exhibited by the exponential probability distribution will play a very important role when we apply this distribution function to service times in a queueing system. Basically, the memoryless property of the exponential probability distribution allows the server to forget what has transpired during the time interval when the server was last busy with a customer or last idle with no customer.

Mathematically, the memoryless property may be expressed as the Markovian probability equation

$$P(T > t + \Delta t \,|\, T > \Delta t) = P(T > t)$$

for any value of t greater than zero and any value of Δt also greater than zero. In other words, the probability of an event occurring in the interval of time $(T, T + t)$ is the same whether there actually was an event occurring at time T or there was no event occurring at time T regardless of how much time Δt has already elapsed. Ross[1] provides a short treatment of the memoryless property as it applies to the exponential probability distribution.

Since we have defined the probability density function of the exponential probability distribution as

$$f(x) = a\exp^{-ax}$$

where parameter a is greater than zero and parameter x is greater than or equal to zero, we may integrate $f(x)$ which will give the corresponding cumulative distribution function for the exponential probability distribution. The cumulative distribution function is:

$$F(x) = \int_0^x a\exp^{-at} dt = 1 - \exp^{-ax} \tag{3-16}$$

Equation 3-16 will hold when the value of parameter x is greater than or equal to zero. The value of $F(x)$ will be equal to zero when the value of parameter x is less than zero.

---

[1] S.M. Ross, *Introduction to Probability Theory Models* (New York: Academic Press, 1972), pp. 109-13.

Furthermore, the mean or expected value and the variance of a random variable
X which has an exponential probability distribution are given as:

$$\text{mean} = \mu = \frac{1}{a} \qquad\qquad (3\text{--}17)$$

$$\text{variance} = \sigma^2 = \mu^2 \qquad\qquad (3\text{--}18)$$

Since the expected value (μ) of a probability distribution is usually the given parameter of the distribution function (rather than the single parameter a for the exponential probability distribution), we could use the following rearrangement to
identify this unknown parameter properly:

$$a = \frac{1}{\mu}$$

In order to generate the required random variates from a given probability distribution, we may apply the inverse transformation method to the given distribution's
cumulative distribution function. In the case of the exponential probability distribution, this is accomplished by taking the distribution's probability density function, given by

$$f(x) = a\exp^{-ax}$$

for values of x greater than or equal to zero and parameter a equal to a constant
greater than zero. The corresponding cumulative distribution function is given by

$$F(x) = 1 - \exp^{-ax}$$

which may be inversed to give

$$ax = -\ln(1 - F(x)) \qquad\qquad (3\text{--}19)$$

where the logarithm is a natural logarithm. Since the exponential probability distribution can be defined in terms of a single parameter (a), and this parameter
may be expressed in terms of the mean (μ) of the distribution, we may make the
following substitution:

$$\mu x = -\ln(1 - F(x))$$

In addition, since $F(x)$ and $1 - F(x)$ are interchangeable because of the symmetry
of the uniform probability distribution, we may also make the following change:

$$\mu x = -\ln(F(x))$$

Since the random numbers which we shall generate from our library function
RND are basically derived from a uniform probability distribution, and since the
range of these generated random numbers are in the range zero to one, we may

also make the following change since the range of the cumulative distribution function is also in the range zero to one:

$$\mu x = -\ln(r)$$

Finally, we may rearrange the above equation in such a manner as to solve for the required parameter which we are looking for, namely:

$$x = -\mu\ln(r)$$

For each generated pseudorandom number from the RND function, a unique value of x may be calculated. The range of this parameter will lie in the interval of the real number line from zero to one, since the natural logarithm of the pseudorandom number r will be less than or equal to zero when the defined range of the pseudorandom r lies in the interval zero to one. The programmed routine which accomplishes this task is listed in Figure 3-11. Gordon[2] and Naylor[3] provide a number of examples and descriptions of the concepts we have just examined.

Gamma Probability Distribution

Our next order of business involves a probability distribution which will be used later to support another probability distribution that has found wide use in queueing theory. Just as the binomial probability distribution supported some aspects of the Poisson probability distribution, the gamma probability distribution will support the Erlang family distributions. An additional feature of the gamma probability distribution is that it also refers to a previously examined probability distribution, namely, the exponential probability distribution. Our primary concern regarding the gamma distribution will involve its relationship with the exponential and the Erlang probability distributions. We will, however, treat this particular

```
1000  INPUT"MU"; MU
1010  FOR N = 1 TO 10
1020    X = -1 * LOG (RND (0.0))
1030    PRINT X
1040  NEXT N
1050  END
```

FIGURE 3-11: Programmed routine to generate $x = -\mu\ln(r)$

[2]G. Gordon, *System Simulation* (Englewood Cliffs, N.J.: Prentice-Hall, 1969), pp. 71-73, 98-101.

[3]T.H. Naylor et al. *Computer Simulation Techniques* (New York: Wiley, 1966), pp. 70-73.

probability distribution as we have treated others in this chapter. Initially, we must present the probability density function (both in a mathematical sense as well as a graphical sense). Also, the corresponding mean value as well as the variance of the distribution must be treated. An appropriate programmed routine will be given which will allow us to present the necessary results as quickly and as accurately as possible.

A gamma probability distribution may be simply defined by two parameters: parameter a and parameter b. These required (and only) parameters will effectively determine the resulting shape of the curve for the probability density function $f(x)$. Equation 3-20 is the equation for the gamma probability distribution with parameters a and b:

$$f(x) = \frac{b^a x^{(a-1)} \exp^{-bx}}{(a-1)!} \qquad (3\text{--}20)$$

The restrictions for the gamma distribution probability density function are that the value for the parameter x must be greater than or equal to zero and the two parameters must have a real value greater than zero. As with the normal probability distribution, the gamma probability distribution does not have an explicit cumulative distribution function. Rather, tables or graphs are usually referenced in order to gain the necessary values for this function.

Although the cumulative distribution function for a gamma probability distribution is not as readily available as most cumulative distribution functions, we may, nevertheless, advance to the important statistical properties of the gamma probability distribution by giving the equations which solve for the mean and the variance. These equations are as follows:

$$\text{mean} = \mu = a/b \qquad (3\text{--}21)$$

$$\text{variance} = \sigma^2 = a/b^2 \qquad (3\text{--}22)$$

As can be seen, the only required input parameters to these equations are the given parameters of the gamma probability distribution, namely, parameters a and b.

As we shall discover when we discuss the Erlang probability distribution, when parameter a takes on the value of a positive integer, the gamma probability distribution will correspond to the Erlangian probability distribution. On the other hand, when parameter a is equal to one, the gamma probability distribution corresponds identically to the exponential probability distribution, that is:

$$f(x) = \frac{b^a x^{(a-1)} \exp^{-bx}}{(a-1)!}$$

If we now set parameter a to one, the above equation will be given as:

$$f(x) = b \exp^{-bx}$$

Replacing parameter b with parameter a yields

$$f(x) = a \exp^{-ax}$$

which is, of course, the probability distribution function for the exponential probability distribution (see Figure 4-3). The corresponding mathematics which show that the gamma probability distribution does, in fact, become the Erlang probability distribution will be treated in the next section on Erlang probability distribution.

We may apply a number of methods to the gamma probability distribution in order to generate the necessary random variates. Although the generation of a random variate for a gamma probability distribution is not as easily accomplished as with other probability distributions, there are a number of techniques which do accomplish this task. Phillips offers a unique solution when the value given for parameter a is strictly a noninteger; his routines also provide for a number of additional statistical properties. Naylor offers the generator solution when the value of parameter a in a probability density function for the gamma probability distribution is strictly an integer. Both solutions offer an effective and efficient technique for solving the problem of random variate generation.

As was mentioned earlier, probability distributions are not usually defined in terms of their respective parameters. Rather, the expected value and the variance of the given probability distribution are usually the given parameters. Therefore, we must have at our disposal the necessary equations which allow us to determine the unknown parameters for any given probability distribution. In this case of the gamma probability distribution, with parameters a and b, the mean and the variance were given as:

$$\mu = a/b$$

$$\sigma^2 = a/b^2$$

This gives the following solutions for parameters a and b:

$$b = \mu/\sigma^2$$

$$a = (\mu)^2/\sigma^2$$

Erlang Probability Distribution

A.K. Erlang, while studying telephone traffic problems in 1917, proposed a family of probability distributions which now bear his name: the Erlang probability distribution functions with parameter k. We shall refer to these probability functions simply as the Erlang-k probability distribution. We shall consider the Erlang-k probability distribution in a later chapter when we examine the probability distributions of certain types of service times within a queueing system. For now, however, let us work on the definition for this particularly important class of probability distributions in terms of the mathematical requirements which we have been following with previous probability distributions, namely, defining the probability density function as well as the calculations for the mean and the variance of the Erlang-k probability distribution.

The Erlang-k probability distribution is often cited as the function which effectively handles the vast majority of cases which fall between the constant probability distribution on the one hand and the exponential probability distribution at the other extreme. In particular, the Erlang-k probability distribution handles the middle ground cases with regard to the service-time requirements of queueing systems where most real-world service-time probability distributions tend to reside.

Our initial equation below gives the probability density function, $f(x)$, for the Erlang-k probability distribution:

$$f(x) = \frac{(\mu k)^k}{(k-1)!} \, x^{(k-1)} \exp^{-(kx\mu)} \tag{3-23}$$

where parameter k must be a positive nonzero integer and the value of x must be greater than or equal to zero. The parameter denoted by the letter mu (μ) must be a positive real number. We shall see later the significance of this particular parameter when we examine the primary characteristics of typical queueing systems in the real world.

In our definition for the gamma probability distribution, we made a reference to the Erlang probability distribution which stated that when parameter k in the Erlang-k probability distribution was an integer, then the probability distribution for the gamma case would be identical to the probability distribution for the Erlang-k probability function. In addition, we may show that when parameter k takes on the value one, the Erlang-1 probability distribution function becomes the exponential probability distribution function. In other words, if we were to take the probability distribution function of the Erlang-k distribution as

$$f(x) = \frac{(\mu k)^k}{(k-1)!} \, x^{(k-1)} \exp^{-(kx\mu)}$$

and we were to set parameter k equal to zero, then

$$f(x) = \mu \exp^{-(\mu x)}$$

where parameter μ (from the Erlang-1 probability density function) would correspond to parameter a of the exponential probability density function. The Erlang-k probability density function may also assume the characteristics of the probability density function of the constant probability distribution when parameter k of the Erlang-k distribution assumes an infinite value. It is within these two extremes, that is, when $k = 1$ and $k = \infty$, that the Erlang-k probability distribution may exhibit the characteristics of a probability distribution which runs the gamut from an almost constant distribution to an almost random distribution. Certainly, there are many situations where these in-between cases must be considered useful.

The programmed routine in Figure 3-12 as well as the resulting output in Figure 3-12 illustrate the various values which the Erlang-k probability density function may assume for various values of parameter k. Figure 3-14 graphs the coordinate points given by the output from the programmed routine.

```
1000  INPUT"MU, K"; MU, K
1010  FOR X = 0 TO 5 STEP 0.5
1020    NM = K - 1
1030    NF = 1
1040    FOR NS = NM TO 1 STEP -1
1050      NF = NF * NS
1060    NEXT NS
1070    EX = (((MU*K)↑K)/NF)*(X↑(K-1))*(EXP(-1*MU*K*X))
1080    PRINT X, EX
1090  NEXT X
1100  GOTO 1000
1110  END
```

FIGURE 3-12: *Programmed routine to generate Erlang variates*

MU, K? 1,2		MU, K? 1,4	
0.0	0.0	0.0	0.0
0.5	0.7358	0.5	0.7218
1.0	0.5413	1.0	0.7815
1.5	0.2987	1.5	0.3569
2.0	0.1465	2.0	0.1145
2.5	0.0674	2.5	0.0303
3.0	0.0297	3.0	0.0071
3.5	0.0128	3.5	0.0015
4.0	0.0054	4.0	0.0003
4.5	0.0022	4.5	0.0001
5.0	0.0009	5.0	0.0000
MU, K? 1,3		MU, K? 2,2	
0.0	0.0	0.0	0.0
0.5	0.7531	0.5	1.0827
1.0	0.6721	1.0	0.2931

FIGURE 3-13: *Output samples from Figure 3-12*

1.5	0.3374	1.5	0.0595
2.0	0.1339	2.0	0.0107
2.5	0.0467	2.5	0.0018
3.0	0.0150	3.0	0.0003
3.5	0.0046	3.5	0.0001
4.0	0.0013	4.0	0.0000
4.5	0.0004	4.5	0.0000
5.0	0.0001	5.0	0.0000

FIGURE 3-13: (Continued)

Values for the mean as well as for the variance of the Erlang-k probability distribution are obtained by the application of the equations

$$\text{mean} = 1/\mu \tag{3-24}$$

$$\text{variance} = (1/(k)^{1/2})(1/\mu) \tag{3-25}$$

where parameter k must be a positive nonzero integer and parameter μ must be positive.

We must wait until we rediscover the Erlang-k probability distribution in a later chapter in order to see how it is actually applied to a real-world situation regarding a waiting-line problem. In particular, we shall use the Erlang-k probability distribution when we model the service-time requirements of certain queueing systems. At that point, reference will be made again to the probability density function as well as to the mean and the variance equations which solve for these particular statistical properties.

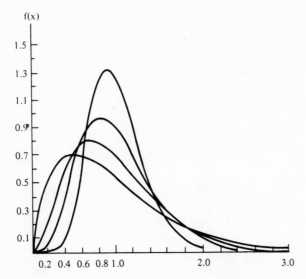

FIGURE 3-14: Graph of typical Erlang-k probability distributions

Queueing System Fundamentals

The fundamental elements of any queueing system involve the customers who demand service from the system and the service facility which provides this service via a server or servers. Basically, a customer arrives at the scene of a queueing system and joins a waiting line. This customer waits in line until service is rendered. If queueing systems were this simple, there would be no pressing need for a further study of the subject. Since this is not the case we must study these real-world queueing systems in light of the particular elements of the system, namely, customers and services.

Figure 4-1 is a summary description of the symbols which we shall use during our study of specific queueing systems. Generally, the notations are those normally utilized in the literature in order to standardize the terminology of queueing theory. The notations used in this book are those recommended in the *Queueing Standardization Conference Report* (May 11, 1971) which was issued by ORSA, AIIE, CORS, and TIMS representatives.

c	Number of identical servers in the service facility
λ	Average customer arrival rate (average number of customers arriving in one unit of time t)
λ'	Traffic rate of arriving customers (equal to λ/c)
m	Number of potential customers in the source population
μ	Average service rate per server (average number of customers served in one unit of time t)
n	Number of customers in the queueing system (total number of customers in the waiting line and the service facility)

FIGURE 4-1: Queueing symbols

ρ Server utilization ration (equal to λ'/μ where $\lambda' \leqslant \lambda$)

L   Average number of customers in the queueing system (that is, customers in the waiting line and the customer in the service facility)

L_q Average number of customers in the waiting line

W   Average time to complete service (includes the time spent in the waiting line)

W_q Average time spent in the waiting line

P(t)   Probability that the time to complete service is greater than the time denoted by t

$P_0(t)$ Probability of finding zero customers in the system at time t

$P_n(t)$ Probability of finding n customers in the system at time t

$P_q(t)$ Probability that the time spent waiting for service to begin is greater than the time denoted by t

FIGURE 4-1: (Continued)

In order to clarify the terminology of queueing systems further, we must also define the terms used to name the various components of a given queueing system. The first of these terms is referred to as a *queueing system*. For all intents and purposes, a queueing system will be considered to be the waiting line or the queue as well as the available service facility which contains one or more servers. The *queue* is known also as the *waiting line*, that is, the queue is the place where customers spend most of their time in the queueing system. In general, we shall consider the term *customer* as any entity demanding or requiring service within a queueing system; a customer, therefore, may be a person waiting in a line at a grocery store checkout counter, a patient in a doctor's office, or an automobile driver waiting in a line at a gasoline station. Likewise, a customer may be the component of a queueing system which takes on the characteristics of an ocean liner waiting for an empty pier, a telephone call waiting for an open trunk line in a switching office, or a broken machine waiting for a repairperson to arrive and fix it. The term *service*, as applied to queueing systems, will refer to that which the customer is demanding of the service facility; for example, service may be thought of as an empty hospital bed, a portion of computer time, or a green light. The remaining component in any queueing system is the *server(s)* who make up the service facility. For our purposes, a server can be either human or machine; for example, a customer waiting in a doctor's office is waiting for service to be rendered by a doctor, while a customer waiting for an open telephone trunk line is waiting for service from an electronic switching device.

Since we shall be considering only the mathematical or the mechanical aspects of a queueing system, it does not matter whether the customer and server(s) are human or machine. To the real-world customer or server(s) it matters tremendously, as witnessed in any real-world queueing system inhabited by human customers or servers. In other words, we shall not concern ourselves with the physiological aspects of waiting lines, but rather with the aspects of queueing systems which lend themselves to mathematical solution.

The following items represent the four major components of any queueing system which we shall consider:

1. A source population of potential customers
2. A particular customer arrival pattern
3. The queue discipline
4. The service mechanism

In addition, there are a number of subcategories related to each major component; for example, a source population may be considered to be finite or infinite in size; a customer arrival pattern, although normally considered in terms of a Poisson probability distribution, may also be considered in terms of any acceptable probability distribution which accurately reflects the pattern of customer arrivals in the real-world queueing system; the queue discipline, as will be seen later, is often considered to be a first-in first-out discipline, but may, for reasons to be dealt with later, be considered otherwise; the service mechanism involves a number of subcategories which measure server utilization, queue length, and so on. Each of these subcategories will be dealt with in greater detail as we are introduced to them in subsequent sections of this chapter. Our initial examination of queueing systems begins with the definitions required for each of the major components and their subcategories.

The Source Population

Customers within a queueing system do not appear magically at a servicing area, although when we are in a hurry customers seem to appear as if they were coming out of the woodwork. The first basic element of a queueing system is called the *source population of potential customers*. This source population of potential customers into the queueing system may be considered to be either finite or infinite in size. For example, a finite source population would consist of the machines on a factory floor which, on occasion, break down and require a repairperson's attention. An infinite source population, on the other hand, may also be considered in terms of being a very large finite source population in addition to being mathematically infinite in size. An example of an infinite source population could be the number of automobiles which require gasoline in their service demand; although the number of automobiles is a finite value, the values is so large that it may, for all intents and purposes, be considered as countably infinite.

The mathematics for the infinite source population are more manageable than are the corresponding calculations for the finite source population. The number of potential customers from a finite source population will have a direct bearing on certain parameters of the queueing system. An infinite source population, on the other hand, does not impact the queueing system in the same manner. As an example, if all the customers from a finite source population were currently waiting in a line, then the arrival of any new customer into the queueing system would be nonexistent. Since we shall soon discover that customer arrival is an important part of queueing system mathematics, the loss of this parameter will have a great impact on queueing system performance and the associated measurement calculations of the queueing system. Therefore, unless otherwise stated, we shall assume that our given source population of potential customers is infinite or, at the very least, finitely large.

Customer Arrival Patterns

Now that we have the basic collection of potential customers for the queueing system, we must consider the element which involves the pattern by which these customers leave the source population and enter the queueing system proper. This *arrival pattern of customers* will be specified in terms of a particular probability distribution function rather than by absolute numbers of customers from the given source population. The customer arrival pattern is a very important queueing system parameter, since it impacts the queueing system quite heavily, as evidenced by the past experiences of those who have stood in a line waiting for their turn at the head of the queue.

The two primary descriptors of customer arrivals involve the probability distribution function of the arriving customers and the probability distribution function of the time interval between successive customer arrivals. These two factors are referred to as the *customer arrival time* and the *customer interarrival time*, respectively. The usual type of customer arrival is represented by generating a single customer from the given source population during some period of time. The probability distribution function which represents this is often that of a Poisson process. When a number of customers are generated from the source population at some period of time, these customer arrivals are considered to be bulk arrivals. Our primary concern will be with the single customer arrival. Bulk arrivals are usually treated as a special class of queueing systems and are described in the literature accordingly; readers interested in this class of queueing systems are referred to the Bibliography for further direction.

Customer arrival patterns are normally described in terms of some probability distribution. Generally, we assume that arriving customers are generated from the source population and enter the queueing system at times t_i (where $i = 0, 1, \ldots, n$), where $t_0 < t_i < \ldots < t_n$. The corresponding random variates $r_k = t_k - t_{k-1}$ (where $k \geq 1$) are labeled interarrival times. The interarrival times correspond to the time interval between successive customer arrivals. Furthermore, we must make the assumption that the r_k random variates are independently and identically distributed in order to satisfy the requirements imposed by probability theory regarding the randomness of these events.

The most commonly used arrival pattern distribution is a Poisson process. The Poisson probability distribution was described earlier as a process which will generate the necessary customers into the system randomly yet at a certain average rate. Most real-world queueing systems are thought of as having a Poisson arrival structure.

In order to model accurately the arrival process of our queueing system by a Poisson probability distribution, the following conditions must be taken into account:

First, a peaking condition must not exist, that is, customer arrivals must be independent of the time during which the system is under study. As an example of a peaking situation, let us consider a movie theater waiting line. In most instances, a Poisson probability distribution does not accurately model the typical movie theater line, since most customer arrivals occur during a five or ten minute period of time prior to the opening of the movie rather than at any other previous time. In other words, if we were to study a movie theater waiting line we would consider the interval of time which did not take into consideration the peaking condition of this particular queueing system. A queueing system may be modeled effectively by a Poisson arrival distribution if only the peak period(s) are considered and if the customer arrivals during this period of time can be shown to exhibit the mathematical qualities of a Poisson probability distribution.

Second, the behavior of arriving customers must be independent of past customer arrivals into the queueing system. When a constant customer arrival distribution is used to model the arrival pattern, this second assumption is invalid since we can determine when the next customer arrival will occur if we know when the most recent customer arrival had occurred. The independence of future events with regard to past events is an example of the memoryless property of a probability distribution. In other words, the probability distribution "forgets" what happened previously and bases its functioning only on the present condition of the system under study. A real-world example of this occurs when we toss a die. The die may take any one of its six possible values completely independent of what values, if any, it had taken prior to this particular toss. Our die-tossing experiment clearly verified this point.

Third, bulk customer arrivals are highly improbable, if not altogether impossible. Customers must arrive singly during any given period of time in order for a Poisson arrival distribution to have effect.

An important parameter of the Poisson probability distribution is that which measures the average number of customer arrivals in a given period of time. This measure, known as the *mean customer arrival rate*, is denoted by the letter lambda (λ). If the unit of time (t) when this mean customer arrival rate is being considered is in, say, minutes, then λ is dimensioned in customers per minute.

If we were interested in knowing the probability of having n customer arrivals in t minutes of time, where customer arrivals are distributed in a Poisson manner, we would use the equation

$$P_n(t) = \frac{(\lambda t)^n}{n!} \, \exp^{-\lambda t}$$

where parameter λ is a postive integer and parameter n corresponds to the integer values 0, 1, . . . customers. A graph of the Poisson arrival distribution for various values of n is illustrated in Figure 4-2.

The only required input parameter to the equation for the Poisson probability distribution function is the value for parameter λ, that is, the average customer arrival rate into this particular queueing system. The assumed value for time t is usually considered to be one unit of time. A programmed routine which will generate the various $P_n(t)$ values is listed in Figure 3-7. The input parameters for this routine require the following values: the average customer arrival rate (lambda), the time period for which the routine will run (time), and the maximum value of customers which will be allowed into the system (maximum n).

The second descriptor of the customer arrival pattern is the measurement of time between successive customer arrivals in terms of a probability distribution which gives the probability of the interarrival time as being less than a particular point in time. If the customer arrival process follows a Poisson probability distri-

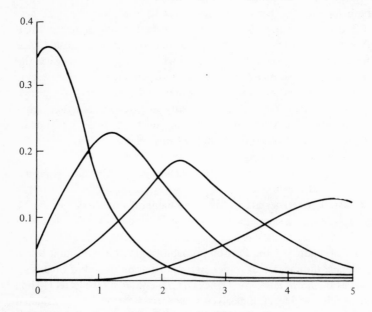

FIGURE 4-2: Graph of the Poisson probability distribution

bution, the corresponding customer interarrival process will follow the exponential probability distribution. Cox and Smith[1] provide proof of this fact. Figure
4-3 illustrates the exponential interarrival distribution derived from the Poisson
arrival distribution of Figure 4-2.

The Queue Discipline

Our introduction to queueing systems so far has shown two of the four major components of a queueing system: the source population of potential customers and
the customer arrival pattern from the source population to the queueing system.
We must now examine the manner in which these arriving customers form into
the waiting line or the queue of the queueing system, that is, the actual organization of the waiting line itself. This description of the queue is known as the *queue
discipline*.

There are two basic aspects of waiting-line organization with which we shall be
concerned. The first relates to the manner in which queued customers will be eventually selected for service. Second, we shall examine the behavior of these queued
customers as they wait in the queue or arrive at the queueing system demanding
their service.

Unless otherwise stated, the manner in which queued customers will be selected
for service will be in the order in which they were queued into the system. This
particular order is referred to as first-in first-out or FIFO. Some texts refer to the
FIFO queue discipline by the acronym FCFS (First-come first-served). The FIFO
queue discipline is the simplest form in so far as the mathematics are concerned
and with regard to its real-world correspondence. For example, customers arriving
at a grocery store checkout counter form a waiting line according to a FIFO queue

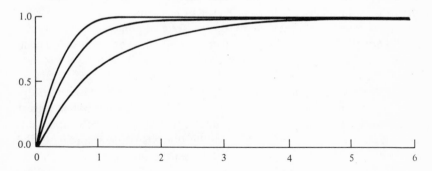

FIGURE 4-3:   Graph of the exponential interarrival times

[1] D.R. Cox and W.L. Smith, *Queues* (New York: Wiley, 1963), pp. 6–12.

discipline. The customer who has waited the longest in the line will be first to receive service, followed by the customer who has waited the next longest, and so on until all customers behind this initial customer have been served. Many other real-world examples of a FIFO queue discipline can be easily seen at traffic lights, doctors' offices, gasoline stations, and so on. Although they may not be readily apparent, these examples do, indeed, follow a FIFO queue discipline.

A last-in first-out queue discipline operates in a manner quite similar to that of a stack structure. This queue discipline is referred to the LIFO or LCFS (last-come first-served). The last customer to arrive at the queueing system and join the waiting line will be the first customer to receive service when it is made available to the waiting customers. A real-world example occurs when an empty elevator stops at the topmost floor of a multistory building. The first customer to enter the empty elevator normally moves to the rear of the elevator cab as the elevator descends to the next lower floor. As passengers enter the elevator cab, the earlier entrants are moved further to the rear. Eventually, the elevator reaches the bottom floor of the building. At this time, the elevator door opens to discharge its passengers. This discharging is the service being performed by the queueing system. The last customer to enter the elevator is pressed against the doors of the elevator cab and is, in fact, the first customer to leave. Passengers who entered the elevator at floors prior to these later customers must wait their turn, that is, the last customers into the elevator leaves first, while the first customer leaves last. A cafeteria tray stack is yet another real-world example of a LIFO queue discipline, although this particular example can be thought of more in physical terms as a stack, rather than as a queue structure.

Priority queue disciplines allow service to be offered to customers depending on their priority in relation to the other waiting customers. For example, the president of a company would, in all probability, receive service in the company cafeteria much quicker than would a relative new hire. This bumping, or "women and children first," queue discipline is quite evident in computer queueing systems, such as in telecommunications systems where there are hot messages, hotter messages, and hottest messages to be received or transmitted. As an example of a computerized queueing system, let us consider a typical on-line terminal system where users are assigned various priority levels of use according to their need to gain access to the mainframe and its resources. These various levels of priorities normally run the gamut from the administrative personnel who usually receive the highest level of priority, since their needs for computer time are particularly important, down through the faculty and, finally, to the students (who, according to the administration and faculty, do not normally need as much of the computer's resources). Within each priority level there must be an embedded queue discipline, since there may be several customers with the same priority level requesting service at the same time; in other words, a separate queue discipline must be specified for individual customers within each of the given priority levels.

Polling is yet another queue discipline often encountered in the real world. This particular discipline offers service to a number of waiting-line systems which feed into the same servicing area. For example, a computerized communications system may scan or poll its attached terminals in order to determine if there are customers or requests for service outstanding. These service requests may be for data input or output depending upon the needs of the terminal user. The particular polling discipline will be based on the number of customers as well as on the priority level of these attached customers. A computer communications controller may offer service more often to the requirements of a real-time terminal than it would to a customer who is merely keying in a computer game. The priority level or pecking order of polled communications systems offer a fertile ground for queueing enthusiasts to consider.

Another queue discipline found quite often in the real world involves a random choice among customers waiting for service in a queueing system. In this sense, random implies that every customer has an equal probability of being selected for service as any other customer in the system. When an impartial teacher calls on members of a class to answer questions, the teacher is, in effect, exhibiting a random queue discipline. Although the students may not be demanding or desiring this type of service, the actions shown by this real-world situation do constitute a queueing system. The acronym for this particular queue discipline is referred to as SIRO (service in random order).

There may be many queue disciplines and, of course, there may be many more variations on the queue disciplines just examined. For example, Chang[2] treats a queue discipline referred to as an interruptible discipline. It is rather interesting since it models a priority queueing system with the added feature that service to a waiting customer may be interrupted if an arriving customer has a higher priority than does the customer currently being served. As can well be expected, the attendant mathematics and physical housekeeping for this queueing system are quite complex.

The behavioral characteristics of the arriving and the waiting customers are best described by their physical representations in the real world. The first of these characteristics is known as balking, which is best described as what an arriving customer does when he or she does not enter the waiting line but leaves the queueing system altogether. This customer, for all intents and purposes, is considered to be a lost customer, since he or she did not join the other waiting customers generated from the same source population of potential customers. The reasons given for customer balking usually involve the fact that the arriving customer did not want to join a long and slow-moving waiting line. Businessmen take note of your long lines at the checkout counters!

[2]W. Chang, "Single Server Queueing Processes in Computing Systems," *IBM Systems Journal* 9 (1970): 42–50.

On the other hand, we have seen the customer who reneges or, after joining a waiting line and waiting for some period of time, leaves the queueing system. This customer, being impatient due to the length of the waiting line or the amount of time already spent there, is also considered to be a lost customer. The balker and the reneger are customers whom the operator of a queueing system must consider in terms of lost profits.

Finally, we often see the customer who jockeys in a queueing system. This type arrives at the queueing system and joins a particular waiting line for some period of time. After waiting in this first line, the customer leaves it to join another waiting line, and so on. This behavior is quite evident in banks and grocery stores where there are multiple waiting lines for arriving customers. Again, impatience is often the reason for this type of customer behavior.

The Service Mechanism

The final aspect of a queueing system, and the reason for which a waiting line forms, is called the service mechanism. A service mechanism is the facility which consists of one or more servers who offer what the customer has waited in line for during some period of time. The assumptions made with regard to the servers are as follows:

1. Each individual server may service only one customer at any one time.
2. If a server is not currently busy (that is, serving a customer) and a customer arrives, that arriving customer will be offered service immediately.
3. At the completion of service to any customer, the next customer available for service is immediately offered service.
4. Service is normally independent of the number of customers waiting in the queue and the number of customers previously served.

These assumptions regard the server or servers as being constantly busy if there are customers in the waiting line. The only time a server will be idle is when there are no customers in the queue or all the previously queued customers are currently being served. Also, the manner in which the service is carried out is not dependent on the number of customers waiting in line or the number of customers already served; if a server were able to see the waiting line leading into that server's service area, he or she might be tempted to speed up or slow down the offered service. Likewise, a server who has been constantly busy from the start of the study may decide to slow down or speed up the service process. Neither of these conditions must exist if the queueing system is to be properly modeled in mathematical terms. In other words, the server or servers are not allowed to "see" the waiting line, rather the server or servers operate in a purely mathematical mode, that is, independent of customer activity.

The required input parameter regarding the service facility is called the mean service rate and is denoted by the letter mu (μ). The amount of time required to service a customer must be independent of the input process (assumption 4 above). Service is performed irregardless of what is currently happening in the queueing system as well as what has previously happened in the queueing system with respect to arriving customers. Special cases have been studied which do allow for a speed-up in service times when the customer arrivals begin to overload the queueing system, however these cases will not be handled in this book. If service times were not adjusted and the maximum allowable queue length was exceeded by arriving customers, subsequent arrivals would be lost simply because there would be no room in the waiting line for them. For our purposes, however, we shall consider only service times for arriving customers that are independent of the input process.

Since we have described the major parameter of the service mechanism (i.e., the mean service rate), we may define the service mechanism in terms of a probability distribution function which has as its average values the previously mentioned mean service rate (μ). Although service times may be considered according to an Erlang, a gamma, or a constant probability distribution function, the usual case is to assume that the service times are exponentially distributed. Just as we used the exponential probability distribution to describe customer interarrival times, we also use it to describe the service time requirements of queued customers. The probability that the service time for any particular customer will be less than some time (t) is defined as:

$$P(T < t) = 1 - \exp^{-\mu t}$$

The exponential probability distribution will be our major distribution function when we examine specific queueing systems.

Types of Queueing Systems

Another important parameter with regard to the service mechanism is the number of servers available to assist the arriving customers. We shall consider three basic cases: the single-server, the multiserver, and the infinite-server service facilities. The single-server service mechanism is the simplest, since there is only one server in the facility. The multiserver service mechanism takes into account some finite number of identical servers operating in a parallel configuration. This service mechanism operates in such a manner that each individual server offers the same type of service to all arriving customers, and each arriving customer is selected for service by the next available (idle) server. The infinite-server service mechanism defines the case where the queueing system provides a server for each and every arriving customer; in other words, as soon as a customer arrives into the queueing system from the source population, that customer is immediately served. If only this were true in the real world!

Just as we think of a customer as an entity (human or machine) requiring service from a queueing system, so, too, must we think of the server or servers as being either human or mechanical. For example, a repairperson or a checkout clerk is obviously a human server. On the other hand, a computer, a traffic light, and an empty operating room are obviously mechanized servers. In addition, we need not have a physical waiting line as is normally considered in most real-world queueing systems (for example, a checkout counter or a toll plaza). For a checkout counter or a toll plaza, the components of the queueing system (i.e., the customers, the servers, and the waiting line) are physically represented quite easily. However, for the machine breakdown example or the computerized communications system, it would be difficult, if not impossible, to envision the physical waiting line or queue, let alone the physical representations for the customers and the servers of the queueing system.

A queueing system, therefore, can be illustrated simply as in Figure 4-4. Each element of the system is represented in relation to its proper setting in the system. Customers requiring service are generated over time from a given source population of potential customers. This source population may be either finite or infinite in size depending upon the type of queueing system we are to examine. These generated customers enter the queueing system which is composed of a waiting line and a service facility. Eventually, customers are selected for service and served depending upon the type of service mechanism given for the queueing system under study. After completion of the service process of the given application, the customer is discharged from the system, where that customer will, no doubt, join a source population which will be used to generate customers for yet another of life's many queues.

The basic queueing system in Figure 4-4 may be considered to be a single-queue single-server queueing system, since there is but one queue and one service facility containing a single server. This type of queueing system represents the simplest and easiest of the ones which we shall examine.

An extension of the single-queue single-server queueing system is shown in Figure 4-5—a schematic representation of a single-queue multiserver queueing system. Arriving customers enter the first queue after they are generated from the source population. Service is rendered at the first service facility. Upon completion of this initial servicing routine, the customer immediately enters the second waiting

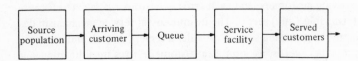

FIGURE 4-4: Single-queue single-server queueing system

line which feeds customers into the second service facility. After service is completed in the second service facility, the customer is discharged into the third queue. Here the customer awaits selection for service in the third servicing area. This entrance and discharge is continued until the customer is finally ejected from the nth service facility where that customer joins a source population consisting of discharged and fully served customers from the single-queue multiserver queueing system.

A real-world example of a single-queue multiserver queueing system may be seen in the operation of a multistation car washing business. Customers (dirty cars) arrive at the initial queue according to some probability distribution from a given source population. These arriving customers immediately enter the first queue. Here they wait in line until their time (selection) for service at the first service facility occurs, say, an overall rinse of the car's exterior. Upon completion of this initial operation, the customer is entered into the second queue and waits for service at the second service facility which may correspond to the soaping of the car. After completion of this second operation, the customer is immediately entered into the next queue to await service at the next service facility. Eventually, the customer enters the last queue, waits for service at the last service facility, and

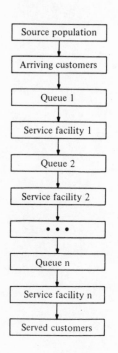

FIGURE 4-5: Single-queue multi-server queueing system

enters the last service area when service is offered. For example, the final rinse, drying, interior cleaning, and so on may be the final phases of this queueing system insofar as the types of service are concerned.

A multiqueue multiserver queueing system (see Figure 4-6) is nothing more than a parallel arrangement of the single-queue multiserver queueing system. In this system, customers generated from the source population are allowed to enter the first queue of queueing system number one, or the first queue of queueing system number two, or the first queue of queueing system number three, or the first queue of queueing system n (where n is equal to the total number of separate single-queue multiserver queueing systems).

A real-world example of a multiqueue multiserver queueing system could be represented as a multistage car wash operation. Each individual system would accept customers according to some means: sports cars to system number one, vans to system number two, eighteen wheelers to system number three, and so on. The acceptance of customers into a particular queueing system could also depend upon

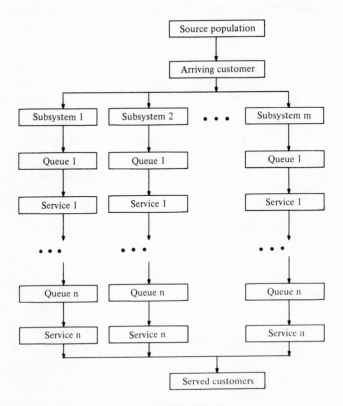

FIGURE 4-6: Multi-queue multi-server queueing system

some other predefined schedule of customer entrance requirements (alphabetically, for example). A multi-island gasoline station which offers full service and self-service pumping areas could also fit into the multiqueue multiserver queueing system configuration. Here customer preference would be the prevailing condition for entrance into any one of the queues. It is also quite possible that there is no preference, that is, customers are entered into any of the multiple queues purely by random choice.

The multiqueue single-server queueing system shown in Figure 4-7 is our last schematic. This waiting-line system can be found at most grocery store checkout counter areas. Arriving customers enter one of *n* queues where they remain waiting until they are selected for service (ringing up their sales) at the facility which is located at the head of the waiting line in which they are queued. Again, customer preference usually is the deciding factor as to which queue an arriving customer will enter from the source population. However, all customers are regarded as entering a queue in a purely random manner. An individual customer may choose one waiting line over the remaining queues; however, on the average, all generated customers behave in a random manner in so far as queue selection is concerned.

In all the above cases, the task of any service facility is similar to the type of service being offered at all the other service facilities within a particular type of queueing system. This is an all important notion which will be considered in-depth in an examination later of the mathematical aspects of queueing systems. Likewise, customer arrival and customer queue selection will be considered as occurring in

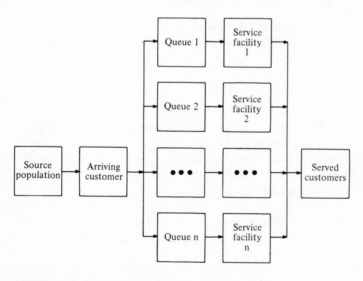

FIGURE 4-7: Multi-queue-single server queueing system

a random manner. It will be seen that this notion of randomness will greatly simplify the mathematics of the queueing systems which we shall encounter.

Queue Measurements

In order to simplify the mathematics of our queueing systems, we must consider these queueing processes as being in a state of statistical equilibrium. Basically, this condition requires that the probability of there being n customers in the system is independent of the time that has passed since the queueing process commenced operation (or observation).

Two principal measurements regarding queueing systems involve the mean number of waiting customers and the mean time these waiting cutomers spend in the system waiting for service. These quantities refer to the total number of customers in the system (those customers waiting for service in the queue or waiting line plus those currently being served) or the quantities may refer to those customers in the queue itself (i.e., only customers currently in the waiting line). These measurements are denoted by the following symbols:

L = mean (average) number of customers in the system

L_q = mean (average) number of customers in the queue

W = mean (average) time to complete service (including the time spent in the waiting line)

W_q= mean (average) time spent waiting in the queue

The following equations, properly interpreted, hold for almost all waiting-line models, assuming the variable λ is a constant:

$$L = \lambda W \tag{4-1}$$

$$L_q = \lambda W_q \tag{4-2}$$

The proof for the above equations in a steady-state queueing system is referred to as Little's formula (see Little[3] and Stidman[4]). In other words, the average number of customers in the queueing system equals the average arrival rate multiplied by the average amount of time a customer spends in the system; similarly, the average number of customers in the queue equals the average arrival rate of customers multiplied by the average amount of the time that a customer spends waiting in the queue.

These relationships are extremely important since they enable all four fundamental quantities of queueing theory to be calculated as soon as any one of them is

[3]J.D.C. Little, "A Proof of the Queueing Formula L = λW," *Operations Research* 9 (1961): 383–87.

[4]S. Stidman, Jr. "A Last Word on L = λW," *Operations Research* 22 (1974): 417–21.

known or calculated by some means. Since quantities are usually easier to determine than others, they will be referenced in our later examination of specific queueing systems.

The traffic intensity ratio is a queueing system measurement which gives the relationship of the average service time for customers (denoted by the symbol λ) to the average time between customer arrivals (i.e., the average interarrival time denoted by μ). This relationship is denoted as:

$$v = \frac{\lambda}{\mu} \qquad (4\text{-}3)$$

The importance of this ratio lies in the fact that it effectively determines the minimum number of servers required in a service facility in order to keep up with the arrival of customers newly generated from the given source population. The unit associated with the traffic intensity ratio is called the Erlang.

In order to gain a feeling for the importance of this ratio, let us examine a typical problem involving the parameters used to calculate this ratio. If we assume that we have a queueing system that exhibits an average interarrival rate (μ) of 10 seconds and an average service rate (λ) of 15 seconds, we would calculate the traffic intensity ratio as 1.5 Erlangs. Since we could not possible have a service facility with one and a half servers, we would apply two servers to the system service facility for the given parameters. A traffic intensity ratio greater than one indicates that the customers are arriving at a faster rate than the single server can assist them. In our example, we would require two servers to handle the rate of arriving customers given the indicated service time of a single-server service facility.

A measurement which affords us the opportunity to examine the server in a queueing system is called the *server utilization ratio*. This particular relationship measures the fractional part of time during which a single server is busy; it is mathematically represented as

$$\rho = \frac{\lambda'}{\mu} \qquad (4\text{-}4)$$

where ρ (rho) is equal to the traffic rate of customers (λ') divided by the system's average service rate. For the single-server queueing system, the traffic rate of customers parameter is equal to the customer arrival rate (λ). However, for the multiserver queueing system, parameter λ' must be altered before it can be applied to equation 4-4: Given c servers in the multiserver queueing system and an average service rate (λ), the traffic rate of customers (λ') will be equal to:

$$\lambda' = \frac{\lambda}{c} \qquad (4\text{-}5)$$

The above equation effectively states that arriving customers are evenly divided among the c servers of a multiserver queueing system service facility.

Since a waiting line normally forms in a real-world queueing system, the quantity given for the traffic rate of customers served by the server (λ') must, of course, be less than or equal to the total arrival rate of customers into the system (λ).

The calculated quantity for the traffic intensity ratio must be less than one, since it is physically impossible for a server to be busy more than 100% of the time. When $\rho = 1$, the server is in a saturated state—100% busy or fully utilized. Therefore, the allowable (theoretical) maximum value which the input rate of customers may assume for a no-loss single-server queueing system is equal to μ.

Allen[5] refers to another measurement employed in queueing theory called the squared coefficient of variation C_x^2 which is defined by the equation

$$C_x^2 = \frac{\sigma_x^2}{(\mu_x)^2} \qquad (4-6)$$

where σ_x^2 is the variance of the random variable X, and μ_x is the mean or expected value of the same random variable X. The particular value assumed by C_x^2 relates to a measurement of the probability distribution for interarrival times and for service times.

When the random variable X is assumed to be a constant, the value attained by C_x^2 will be equal to zero. If the random variable X has an exponential probability distribution, then $C_x^2 = 1$. Furthermore, Allen states that the value attained by the measurement C_x^2 for the Erlang-k probability distribution will be equal to the reciprocal of the parameter k given in the Erlang-k distribution (i.e., $C_x^2 = 1/k$ for an Erlang-k probability distribution). In general, if C_x^2 is almost equal to zero (i.e., if it is very small), then a regular probability distribution defines the arrival process. On the other hand, if C_x^2 is almost one, then the arrival process is considered to be almost a random probability distribution function. The following represents the ranges which C_x^2 may assume and the particular distribution which defines the arrival process (see Allen):

$0.0 < C_x^2 < 0.7$ arrivals tend to be evenly spaced

$0.7 < C_x^2 < 1.3$ arrivals close to Poissonian

$1.3 < C_x^2$ arrivals tend to cluster

In other words, as the value of C_x^2 increases, the congestion tends to increase also, that is, the line becomes longer.

The probability notation $P_n(t)$ refers to the probability that n customers are currently in the queueing system at time t. Although it would seem that the value of $P_n(t)$ would depend only on the given time t, the initial condition of the system

[5]A.O. Allen, "Elements of Queueing Theory for System Design," *IBM Systems Journal* 14 (1975): 163.

must also be taken into consideration. The initial condition of a queueing system involves the number of customers in the system (i.e., the current length of the waiting line) when the service facility opens for business. Normally, there would be zero customers in the queue, however it is entirely possible that there may be n customers waiting in the queue before the service facility begins selecting queued customers. For example, how many times have you arrived at a department store having a fantastic sale only to find a line of customers waiting to get into the store before the doors open? How many times have you driven up to a gasoline station prior to its opening only to find a long line of cars?

The probability of having zero customers in the system at time t is also of extreme importance. This probability notation is given as $P_0(t)$. The value taken by this parameter is quite useful when server idle time is being considered, that is, if there are no customers in the system (waiting for service or currently being served), then the server is 100% idle and the cost of maintaining that idle server is, indeed, a significant contribution to the total cost of the queueing system.

The specific calculations and required input parameters for each of the queue measurements previously defined will be considered again when we deal with specific queueing systems in a later chapter.

Queueing System Shorthand

Kendall[6] proposed a convenient classification system which relates the primary components of a queueing system in an easy to remember shorthand technique. The major parts of all queueing systems can be considered in terms of an input process, the service distribution, and the number of servers. Using the symbols listed in Figure 4-8, we may easily write the description of a queueing system in a shorter manner than if we had to spell out each individual component. The symbols are applied to

$$A/B/c \qquad\qquad (4-7)$$

where parameter A represents the arrival pattern of customers from the given source population. Parameter B represents the service time pattern of the given service facility, while parameter c represents the number of servers in the same service facility. For example, if we were to consider a queueing system which had a Poisson arrival process, an exponential service distribution, and a single server, the Kendall notation for this queueing system would be given as: fn

$$M/M/1$$

[6]D.G. Kendall, "Some Problems in the Theory of Queues," *Journal of the Royal Statistical Society, Series B* 13 (1951): 151–85.

Symbol	Description
M	Poisson arrival process or exponential service times and exponential interarrival time
E_k	Erlangian distribution (of order k) for interarrival or service times
G	General service times
GI	General independent interarrival times
D	Deterministic or constant interarrival or service times

FIGURE 4-8: *Kendall notation elements*

A further extension of Kendall's notation takes into account a number of other parameters required by a given queueing system. The extended notation is given by

$$A/B/c/K/m/Z$$

where the symbols A, B, and c still represent the input process, the service process, and the number of servers, respectively. The additional symbols have the following descriptions:

$K =$ capacity of the system (maximum queue length)

$m =$ number of potential customers in the given source population

$Z =$ type of queue discipline

As an example, an $M/M/1/10/\infty/FIFO$ notation would represent a queueing system with a Poisson arrival distribution, an exponential service distribution, a single server, a maximum allowable queue length of ten customers, an infinite source population of potential customers, and a first-in first-out queue discipline.

Since we shall normally assume a FIFO queue discipline and an infinite source population for many of our queueing systems, the shorter version of Kendall's notation will be used. When necessary, the queueing system will be described fully as well as in terms of the required notational symbols.

Problems for Reader Solution

Define the following terms as they apply to queueing theory:

(a) Average customer arrival rate

(b) Average number of customers in the system

(c) Average queue length

(d) Average service rate

(e) Average time required to complete service

(f) Average time spent in the queue

(g) Customer

(h) Exponential service distribution

(i) First-in first-served

(j) Infinite-server queueing system

(k) Kendall's notation

(l) Multiqueue multiserver queueing system

(m) Multiqueue single-server queueing system

(n) Poisson arrival process

(o) Probability of n customers in the system

(p) Queue

(q) Queue discipline

(r) Server

(s) Server utilization ratio

(t) Service facility

(u) Single-queue multiserver queueing system

(v) Single-queue single-server queueing system

(w) Source population

(x) Traffic intensity ratio

5

Basic Queueing Systems

This chapter will introduce four basic queueing systems denoted by their Kendall notation as M/M/1, M/M/c, M/M/∞, and $M/E_k/1$. The single-server, Poisson input process, and exponential service-time distribution (M/M/1) queueing system is the simplest of the four. It will afford us the opportunity to tie together most of the concepts of probability theory examined previously. In addition, the fundamental concepts of queueing theory will be examined in terms of a real-world queueing system which follows the notation given for this queueing system. The Poisson input process, exponential service-time distribution, and multiserver (M/M/c) queueing system will extend our initial examination of queueing systems by introducing a more difficult system. The Poisson input, exponential service-time, and infinite-server (M/M/∞) queueing system will be examined primarily in terms of its theoretical and mathematical aspects. Our last basic queueing system will be the Poisson input distribution, Erlangian service-time distribution, single-server $(M/E_k/1)$ queueing system. It is included in this chapter because it incorporates a great deal of the material we have previously examined in terms of probability theory as well as fundamental queueing theory. In addition, the $M/E_k/1$ queueing system will allow us to examine a particularly interesting nonexponentially distributed service-time which will lay the foundation for other such queueing systems in chapter 6.

The M/M/1 Queueing System

Our first queueing system will focus on the description of the model which has the characteristics given below taken from the Kendall notation of the M/M/1 queueing system.

The input process will be given a Poisson probability distribution where customer arrivals will be assumed to follow a random occurrence but at a certain average rate. This average customer arrival rate will be denoted by the symbol λ and is dimensioned in customers per unit of time t.

The servicing process will follow a (negative) exponential probability distribution where the service-time distribution is symbolized by the constant average rate given by $1/\mu$. The dimension associated with this average service rate is in customers per unit of time t.

The number of available servers in the service facility is limited to a single server who has an average service rate equal to parameter μ customers served per unit of time t.

The source population from which potential customers will be generated is assumed to be infinite in size. Kleinrock[1] and Allen[2] examine the more complex case where the source population of the M/M/1 queueing system is finite in size.

The queue discipline is of a first-in first-out (FIFO) nature. This is quite common for the M/M/1 queueing system as well as most of the other queueing systems which we shall eventually examine. A last-in first-out (i.e., a stack-type queue discipline) or one of the priority queue disciplines will not be covered at this time.

The allowable maximum length of the waiting line will be assumed as infinite in size, that is, all possible arriving customers from our infinite source population will be allowed entry into the system. Kleinrock examines the finite queue storage situation.

The M/M/1 queueing system is considered to be the simplest of the many queueing systems which are studied in the literature as well as in the real world. The primary reason for its wide popularity lies in the fact that the probability distributions which describe the input process and the service process have a rather simple mathematical form. In addition, the M/M/1 queueing system offers a realistic approach to a queueing system often found in the real world. Customer arrival patterns in a real-world queueing system quite often follow a Poisson probability distribution; therefore, the M/M/1 model can be used as an accurate and effective mathematical tool in determining the various measurement procedures and parameters of many actual queueing systems. For example, the Poisson probability distribution is often used as the mathematical tool in describing aircraft arrivals into an airport and telephone calls into a switchboard. Since these customer arrivals (airplanes and telephone calls) often occur at a certain average rate over a given period of time, the Poisson probability distribution can statistically model these types of customer behavior in a queueing system. Although it can be argued that airplane arrivals are based on a predetermined schedule of deterministic events,

[1] L. Kleinrock, *Queueing Systems, Volume I: Theory* (New York: Wiley, 1975), pp.103–05.

[2] A.O. Allen, "Queueing Models of Computer Systems," *Computer* 13 (1980): 15–20.

there are certain events outside the control of the schedulers which effectively insert the necessary degree of randomness into the behavior of these arrivals which afford the opportunity to use the Poisson probability distribution. For example, start-up problems at the departing terminal, bad weather during the flight, and unanticipated small craft congestion at the destination terminal are some of the uncontrollable (random) events which lend a hand in the accurate modeling of these types of arrivals to the Poisson probability distribution.

The Poisson assumption concerning the input process of the M/M/1 queueing system simply states that the probability of the next occurrence of a customer arrival into the system is independent of the time that has transpired since the last customer arrival into the system. In other words, if Δt represents a small interval of time t and parameter λ represents the average customer arrival rate, then the probability of a customer arrival in the interval of time $(t, t + \Delta t)$ will be equal to the product $\lambda \Delta t$. The probability distribution of customer arrivals given by this particular assumption is called Poisson, since the probability of n such customer arrivals in any finite interval of time t is given by

$$P_n(t) = \frac{\exp^{-\lambda t}(\lambda t)^n}{n!} \qquad (5\text{-}1)$$

where parameter λ represents the average customer arrival rate and parameter λt is called the parameter of the given Poisson probability distribution. Usually, the parameter of the Poisson probability distribution is simply given by parameter λ.

The mean and the variance of the Poisson probability distribution are given by the following:

$$\text{mean} = \sum_{n=0}^{\infty} nP_n(t) = \lambda t \qquad (5\text{-}2)$$

$$\text{variance} = \sum_{n=0}^{\infty} n^2 P_n(t) - (\lambda t)^2 = \lambda t \qquad (5\text{-}3)$$

In addition to noting that the mean and the variance of a Poisson probability distribution with parameter λt are, in fact, equal to λt, it should also be noted that the average rate of arriving customers must be equal to the average rate of departing customers. This is due to the fact that no customers are lost to the system under conditions of stability.

The service process of the M/M/1 queueing system implies that the length of time required to serve a customer fully can be mathematically modeled by the exponential probability distribution, that is,

$$f(x) = 1 - \exp^{-(\mu t)}$$

where parameter μ is the average service rate of the single server. The mean and the variance for the exponential probability distribution are given by the following:

$$\text{mean} = 1/\mu \qquad\qquad (5\text{-}4)$$

$$\text{variance} = 1/(\mu)^2 \qquad\qquad (5\text{-}5)$$

Since the exponential probability distribution essentially "forgets" its past, a service process which follows an exponential distribution is immune to the occurrences of its past. Therefore, the future state of the service is based only on the present condition of the service.

Before we continue with the other required formulas necessary in our examination of the M/M/1 queueing system, we must first review the queueing measurement known as the traffic intensity ratio. We must assume that the average customer arrival rate (λ) is less than the average service rate (μ) in order for equations 5-6 through 5-12 to be true. The ratio of λ to μ must also, therefore, be less than one. This ratio

$$\rho = \lambda/\mu \qquad\qquad (5\text{-}6)$$

is called the *traffic intensity ratio*. Replacement of this ratio (λ/μ) by its designated symbol (ρ) in the equations which follow will allow us to handle the mechanics of the equations more easily. Therefore, the traffic intensity ratio will be calculated initially, since the values for the average customer arrival rate as well as the average service rate must be given input parameters for the M/M/1 queueing system.

Calculations concerned with queue size (i.e., number of customers in the queueing system at time t) involve the application of differential-difference equations, which are treated thoroughly in the probability theory texts listed in the Bibliography. The basic equations developed from this technique will be presented here in a workable form.

Our initial question regarding the M/M/1 queueing system is whether there are zero customers in the waiting line when the system begins operation or after the system has been in operation for some period of time. This particular case is, of course, quite favorable to the lucky customer arriving into the system, for it means that this customer will receive immediate attention at the service facility from the idle server. The only time this customer will spend in the queueing system will be the time required to receive service.

If λ represents the average customer arrival rate and μ symbolizes the average service rate of our single server in the service facility, then the probability of having zero customers in the queueing system at time t is given by the following:

$$P_0(t) = 1 - (\lambda/\mu) = 1 - \rho \qquad\qquad (5\text{-}7)$$

The above equation is often considered to be a special case for any of the queueing systems which we shall examine. A more important equation involves the calculation of the probability that the queueing system will have, at time t, n customers in the system (i.e., $n - 1$ customers waiting in the queue and one customer being served). Again, the traffic intensity ratio (ρ) must be less than one and the average customer arrival rate (λ) must be less that the average service rate (μ). The equation which solves for the probability of having exactly n customers in the queueing system at time t is, therefore, as below:

$$P_n(t) = \rho^n P_0(t) = \rho^n (1 - \rho) \qquad (5\text{-}8)$$

The equations which solve for zero customers in the queueing system at time t as well as the probability of n customers in the system at time t will be used in our programmed routine for the M/M/1 queueing system.

Queueing theory does not only include the solution and analysis of probabilities such as the two previous equations. Other probabilities, such as those which involve customer waiting time and the number of customers in the system during the observation period, are of equal importance. The ease with which the necessary computations for these probabilities can be made with regard to the M/M/1 queueing system with an infinite source population will serve as the vehicle for determining these next important equations.

The most striking aspect of a physical queueing system, at first glance, is the number of customers already waiting for service to be offered. No doubt, your initial reaction to a queueing system in the real world is to scan the available waiting lines and then join the line containing the fewest customers (or the faster moving line). Since we are currently examining the single-server queueing system, the matter of a choice among many queues will be considered in a later section. Unfortunately, the M/M/1 queueing system accepts customers only into a single waiting line; therefore, the equations which follow will assist us in determining the necessary values for measuring the effectiveness of this particular queueing system. The equations we soon shall examine involve the average number of customers in the queueing system during our observation period. These waiting customers must be considered in terms of the customers in the waiting line as well as the customer in the service facility who is currently being served by the single server. We shall consider the waiting customers (i.e., those in the waiting line) as being the customers in the queue. The single customer currently being served will be considered as the customer in the service facility.

Since the number of customers who will wait in line is determined by the number of arriving customers and the service rate of those customers in the service facility, the equation used to calculate the average number of customers in the system involves the input parameters previously defined, that is, the average customer arrival rate (λ) and the average service rate (μ). The general equation which will calculate the average number of customers in the queueing system, that is, customers waiting in line as well as those currently being served is given as follows:

$$L = \frac{\lambda}{\mu - \lambda} \qquad (5\text{-}9)$$

Since the length of the waiting line, that is, those customers waiting in line prior to being selected for service, is also important, the equation which calculates the average queue length is as follows:

$$L_q = \frac{\lambda^2}{\mu(\mu - \lambda)} \tag{5-10}$$

Of course, once a customer arrives into the queueing system from the source population and enters the waiting line, the next question is, how long will that customer have to wait in the line before receiving service? Since this waiting line is dependent upon the number of customers waiting ahead of this newly arrived customer, the value calculated for L_q must be taken into consideration. Therefore, the average waiting time of a customer prior to being selected for service in our M/M/1 queueing system is as below:

$$W_q = \frac{\lambda}{\mu(\mu - \lambda)} \tag{5-11}$$

After a customer waits in the queue for W_q units of time (on the average, of course), the customer must now spend a certain amount of time being served in the single-server service facility. This period of time will be considered to be the average service time. However, we must now define the equation which will calculate the amount of time the customer spends in the queueing system (i.e., waiting in line and receiving service):

$$W = \frac{1}{\mu - \lambda} \tag{5-12}$$

As an example of the application of these equations for the single-server single-waiting line infinite population queueing system, let us consider a typical real-world situation—a one-pump gasoline station. Let us also assume that the source population of potential customers which will generate customers over time is infinite or, at the very least, large enough to warrant it being called infinite. We assume that customers arrive according to a Poisson probability distribution; this is the usual case even in a typical real-world situation and is not being assumed for its mathematical simplicity. Our example will consider the particular case where the average customer arrival rate equals five customers per each one-hour unit of time. Service will, of course, follow an exponential probability distribution with an average service rate of ten customers per hour. Our input parameters, therefore, are as follows:

$$\lambda = 5 \text{ customers per hour}$$

$$\mu = 10 \text{ customers per hour}$$

Hence, the traffic intensity ratio is as follows:

$$\rho = \lambda/\mu = 5/10 = 0.5$$

This indeed, satisfies the restriction that the traffic intensity ratio must be less than one.

Applying the above input parameters to our equations which solve for the average number of customers in the system and the waiting time of these average customers, we find the following:

$$L = \frac{\lambda}{\mu - \lambda} = \frac{5}{10 - 5} = 1.0 \text{ customers waiting in the system}$$

$$L_q = \frac{\lambda^2}{\mu - \lambda} = \frac{5^2}{10 - 5} = 0.5 \text{ customers in the queue}$$

$$W_q = \frac{\lambda}{\mu(\mu - \lambda)} = \frac{5}{10(10 - 5)} = 0.1 \text{ hours spent in the queue}$$

$$W = \frac{1}{\mu - \lambda} = \frac{1}{10 - 5} = 0.2 \text{ hours waiting in the system}$$

The probability that there will be zero customers in the system, given our previous input parameters, is given by the equation below:

$$P_0(t) = 1 - (\lambda/\mu) = 1 - (1/2) = 0.5$$

Solving for n customers in the system at time t (for values of $n = 1, 2, 3,$ and 4) follows by applying the general formula

$$P_n(t) = \rho^n(1 - \rho)$$

that is,

$$P_1(t) = 0.5^1(1 - 0.5) = 0.25000$$
$$P_2(t) = 0.5^2(1 - 0.5) = 0.12500$$
$$P_3(t) = 0.5^3(1 - 0.5) = 0.06250$$
$$P_4(t) = 0.5^4(1 - 0.5) = 0.03125$$

The programmed routine listed in Figure 5-1 is our computerized model of an M/M/1 queueing system. The required input parameters to this routine include the average customer arrival rate (λ) and the average service rate (μ). Output includes the answers to the various equations examined in this section. Figure 5-2 is some typical output for the input parameters given.

```
1000  REM    M/M/1 QUEUEING SYSTEM
1010  REM    VARIABLE-NAME    DESCRIPTION
1020  REM       L      AVERAGE NUMBER OF CUSTOMERS IN SYSTEM
1030  REM       LM     LAMBDA - AVERAGE CUSTOMER ARRIVAL RATE
1040  REM       LQ     AVERAGE NUMBER OF CUSTOMERS IN QUEUE
1050  REM       MU     MU - AVERAGE SERVICE RATE
1060  REM       N      CUSTOMER COUNTER
1070  REM       P(0)   PROBABILITY OF 0 CUSTOMERS IN SYSTEM
1080  REM       P(N)   PROBABILITY OF N CUSTOMERS IN SYSTEM
1090  REM       RO     RHO - TRAFFIC INTENSITY RATIO
1100  REM       W      AVERAGE WAITING-TIME IN SYSTEM
1110  REM       WQ     AVERAGE WAITING-TIME IN QUEUE
1120  DIM  P(5)
1130  PRINT"ENTER AVERAGE CUSTOMER ARRIVAL RATE (LAMBDA)";
1140  INPUT LM
1150  PRINT"ENTER AVERAGE SERVICE RATE (MU)";
1160  INPUT MU
1170  IF LM/MU <= 1.0 THEN 1240
1180    PRINT"TRAFFIC INTENSITY RATIO (LAMBDA/MU) MUST BE"
1190    PRINT"LESS THAN ONE IF THE PROBABILITY OF N CUSTOMERS"
1200    PRINT"IN THE SYSTEM AT TIME T IS TO BE INDEPENDENT"
1210    PRINT"OF THE TIME T."
1220    PRINT
1230    GOTO 1130
1240  RO = LM/MU
1250  P(0) = 1.0 -RO
1260  FOR N = 1 TO 5
1270    P(N) = (RO↑N) * (1.0-RO)
1280  NEXT N
1290  L = LM/(MU - LM)
1300  LQ = (LM↑2) / (MU * (MU - LM))
1310  W = 1.0/(MU - LM)
1320  WQ = LM/(MU * (MU - LM))
1330  PRINT"PROBABILITY (P) OF (N) CUSTOMERS IN THE SYSTEM"
1340  PRINT"(N)            (P)"
1350  FOR N = 0 TO 5
1360    PRINT N, P(N)
1370  NEXT N
1380  PRINT"AVERAGE NUMBER OF CUSTOMERS IN THE SYSTEM ="; L
1390  PRINT"AVERAGE NUMBER OF CUSTOMERS IN THE QUEUE ="; LQ
```

FIGURE 5-1: Programmed routine for the M/M/1 queueing system

1400 PRINT"AVERAGE WAITING–TIME IN THE SYSTEM ="; W
1410 PRINT"AVERAGE WAITING–TIME IN THE QUEUE ="; WQ
1420 GOTO 1130
1430 END

FIGURE 5-1: (Continued)

ENTER AVERAGE CUSTOMER ARRIVAL RATE (LAMBDA)? 5
ENTER AVERAGE SERVICE RATE (MU)? 10

PROBABILITY (P) OF (N) CUSTOMERS IN THE SYSTEM
(N)	(P)
0	.500
1	.250
2	.125
3	.063
4	.031
5	.016

AVERAGE NUMBER OF CUSTOMERS IN THE SYSTEM = 1.0
AVERAGE NUMBER OF CUSTOMERS IN THE QUEUE = 0.5
AVERAGE WAITING–TIME IN THE SYSTEM = 0.2
AVERAGE WAITING–TIME IN THE QUEUE = 0.1

ENTER AVERAGE CUSTOMER ARRIVAL RATE (LAMBDA)? 9
ENTER AVERAGE SERVICE RATE (MU)? 10

PROBABILITY (P) OF (N) CUSTOMERS IN THE SYSTEM
(N)	(P)
0	.10
1	.09
2	.08
3	.07
4	.07
5	.06

AVERAGE NUMBER OF CUSTOMERS IN THE SYSTEM = 9.0
AVERAGE NUMBER OF CUSTOMERS IN THE QUEUE = 8.1
AVERAGE WAITING–TIME IN THE SYSTEM = 1.0
AVERAGE WAITING–TIME IN THE QUEUE = 0.9

FIGURE 5-2: Sample output from programmed routine of Figure 5-1

ENTER AVERAGE CUSTOMER ARRIVAL RATE (LAMBDA)? 1
ENTER AVERAGE SERVICE RATE (MU)? 10

PROBABILITY (P) OF (N) CUSTOMERS IN THE SYSTEM

(N)	(P)
0	.90
1	.09
2	.01
3	.00
4	.00
5	.00

AVERAGE NUMBER OF CUSTOMERS IN THE SYSTEM = 0.11
AVERAGE NUMBER OF CUSTOMERS IN THE QUEUE = 0.01
AVERAGE WAITING-TIME IN THE SYSTEM = .11
AVERAGE WAITING-TIME IN THE QUEUE = .01

ENTER AVERAGE CUSTOMER ARRIVAL RATE (LAMBDA)? 3
ENTER AVERAGE SERVICE RATE (MU)? 10

PROBABILITY (P) OF (N) CUSTOMERS IN THE SYSTEM

(N)	(P)
0	.70
1	.21
2	.06
3	.02
4	.01
5	.00

AVERAGE NUMBER OF CUSTOMERS IN THE SYSTEM = 0.429
AVERAGE NUMBER OF CUSTOMERS IN THE QUEUE = 0.129
AVERAGE WAITING-TIME IN THE SYSTEM = 0.143
AVERAGE WAITING-TIME IN THE QUEUE = 0.043

FIGURE 5-2: (Continued)

We shall now turn our attention to a typical example problem. Let us assume that customers arrive at a telephone booth in a Poisson manner with an average interarrival rate equal to 10 minutes, that is, the average time between successive customer arrivals is 10 minutes. Therefore, parameter λ will be equal to 0.1 customer arrivals per minute. The length of an average telephone call is exponentially distributed with an average service time of three minutes, that is, there will be 0.33 service activities per minute ($\mu = 0.33$).

Our obvious first question asks, what is the probability that an arriving customer will have to wait for service? The solution is found by the following calculation:

P (arriving customer must wait) $= 1 - P_0$

$$= \lambda / \mu$$

$$= 0.1/0.33$$

$$= 0.30$$

There is a one in three chance that an arriving customer must wait for service to
be offered, that is, for the telephone booth to become empty.

The second question asks, what is the average waiting-line length, or how many
customers are (on the average) waiting for service? The following calcualtion solves
this problem:

Average waiting-line length $= \dfrac{\mu}{\mu - \lambda}$

$$= \dfrac{0.33}{0.33 - 0.10}$$

$$= 1.43 \text{ customers in the queue}$$

Since an arriving customer will (on the average) find 1.43 customers ahead of him
or her upon arriving at a telephone booth, the telephone company has decided to
install another booth near this one only when an arriving customer finds a waiting
line time of three minutes or longer. Our last question for this problem asks, by
how much does the arrival flow of customers have to increase in order to justify
this additional telephone booth? The following solves this problem:

Average wait for service $= \dfrac{\lambda}{\mu(\mu - \lambda)}$

$$3 = \dfrac{\lambda}{0.33(0.33 - \lambda)}$$

$$\lambda = 0.16 \text{ arriving customers per minute}$$

The customer arrival flow must increase from six to 10 arrivals per hour in order
to justify the addition of a second telephone booth.

Our next example problem involves an M/M/1 queueing system with the following parameters:

Average customer arrival rate $(\lambda) = 0.1$ arrivals per minute

Average service rate $(\mu) = 0.33$ customers served per minute

The first question we would like answered is, what is the probability that an arriving customer will have to wait more than 10 minutes before service is offered? Our solution is below:

$$P \text{ (waiting time} \geqslant 10) = \int_{10}^{\infty} (1 - \rho) \, \lambda \exp^{(\lambda - \mu)w} dw$$

$$= \rho \exp^{(\lambda - \mu)w} \Big]_{\infty}^{10}$$

$$= 0.3 \exp^{-2.3}$$

$$= 0.03$$

After the customer is offered service, the next question involves the total time which this customer spends in the waiting line and the service facility: What is the probability that this customer will spend more than 10 minutes waiting and being served? The solution is:

$$P \text{ (total time in the system} \geqslant 10) = \int_{10}^{\infty} (\mu - \lambda) \exp^{(\lambda - \mu)v} dv$$

$$= \exp^{10(\lambda - \mu)}$$

$$= \exp^{-2.3}$$

$$= 0.10$$

Since most queueing system problems involve cost considerations, we shall examine a typical real-world queueing problem which takes into account the cost of operating a queueing system. In particular, let us assume that we must decide between one of two repairpersons which we have on our staff. The reason for having a repairperson is that we have a number of machines which break down at an average rate of three machines per hour. It may be further assumed that the machines break down in a Poisson manner. When one of these machines break down, the nonproductive time costs our company $5 per hour. Our choice of a repairperson will be determined by which of the two is the least costly based on the following characteristics: Repairperson number one has an average fix rate of four machines per hour and costs $3 per hour. The second repairperson costs $5 per hour but has an average fix rate of six machines per hour. Which of the two repairpersons is the "better"?

The salary for the first repairperson is:

Salary = $3 per hour x 8 hours per day, or $24 per day

The second repairperson's salary is calculated as:

Salary = $5 per hour x 8 hours per day, or $40 per day

At this point in our calculations we would assume that the first repairperson would be less costly than would the second. However, the "better" repairperson is to be determined not only on salary but weighted by his or her average nonproductive time based on the repairperson's fix rate of broken machines. The average nonproductive time of the first repairperson is calculated by:

$$\text{Average nonproductive time} = 8\,\frac{\lambda}{\mu-\lambda}$$

$$= 8\,\frac{3}{4-3}$$

$$= 24 \text{ machine hours}$$

The nonproductive time is now multiplied by the repairperson's hourly rate of $5, yielding the average nonproductive value of $120.

The second repairperson has an average nonproductive time of:

$$\text{Average nonproductive time} = 8\,\frac{3}{6-3}$$

$$= 8 \text{ machine-hours}$$

The average nonproductive cost (eight machine-hours multiplied by $5 per hour) gives a value of $40.

We now add the salary costs and nonproductive costs of each of the two repairpersons in order to find the "better" of the two:

First repairperson = $120 + $24 = $144

Second repairperson = $40 + $40 = $80

Therefore, the second repairperson is the "better" of the two since the total cost difference is $144 - $80 = $64 less.

The administration of a university computing center has decided to offer a single on-line terminal to students as an alternative data-entry device. This decision was brought about by two factors: to ease the load on the batch-processing system and to acquaint the students with the operation of an on-line device for program and data entry. The terminal will initially be offered during normal class hours for an eight-hour period. Student feedback is, of course, desired by the administration as to the overall effect which this device may have on their usage. After a number of weeks in operation, the computing center administration studies the comments of the students and is surprised to hear that the students complain of long waits for access to the terminal.

Statistics have been gathered which show the following characteristics of this initial usage period. The arrival of students (customers) to the terminal follows a Poisson pattern with an average customer arrival rate (λ) equal to 20 customers per eight-hour period (2.5 customers per hour). The average time spent using the terminal (service time) is equal to 20 minutes (three customers per hour are served, on the average). This service rate is found to follow an exponential distribution.

Students are served on a first-come first-served basis. Although the students form a finite source population, their number is so large that the source population may be considered to be infinite.

Given the previous characteristics, it may be assumed that we have an M/M/1 queueing system with the following input parameters: $\lambda = 2.5$ and $\mu = 3$. Initially, we may calculate the facility utilization at 83.3% by the following:

(20 customers per each eight hours) x (0.333 hour per customer)

In other words, during a typical eight-hour usage period, the terminal is in actual use for six hours and 40 minutes (rather good, considering the conditions). Given this initial calculation and the fact that the students are complaining of long lines to the terminal, we may ask the question, are the students' complaints justified?

We may apply the equations for the M/M/1 queueing system in order to arrive at a numerical answer to this question. Equations 5-7 through 5-12 will be the basis for these calculations, while the programmed routine in Figure 5-1 will accomplish the mathematical manipulations required by this question. The following "answers" supply an initial estimate of the degree of justified complaint level:

$$P_0 = 0.1667$$
$$P_1 = 0.1389$$
$$P_2 = 0.1157$$
$$P_3 = 0.0965$$
$$P_4 = 0.0804$$
$$P_5 = 0.0670$$

The probability of having n customers in the system is not very high. For example, there is approximately a 14% chance that there will be a single customer in the system and that customer will be currently using the terminal. Having to wait for two customers to complete service is about 12%. Perhaps if you were the student desiring to use the terminal these probabilities might prove to be excessive.

In addition, the following queueing parameters are given from our study of this system:

$$L = 5.0000$$
$$L_q = 4.1667$$
$$W = 2.0000$$
$$W_q = 1.6667$$

The average length of the line (including the customer using the terminal) is equal to five students. The average length of time which a typical customer would be required to wait before gaining access to the terminal is equal to 1.667 hours (one hour and 40 minutes). If the average customer uses the terminal for 20 minutes, then that customer is expected to wait for one hour and 40 minutes. It seems that the students' complaints are justified considering the time they must wait to use the terminal. Would you wait that long in order to have a service period of 20 minutes?

Before the computing center administration can reach a decision regarding the acquisition of one or more added terminals, the final days of the semester arrive with the following customer activities. Customer arrivals increase from the pre-final days value of 20 customer arrivals per eight-hour period to 30 customer arrivals per eight hours, that is, the hourly customer arrival rate is now 3.75 distributed in a Poisson manner. Since most of the students have the bulk of their programs written, service is required only to debug the programs or to obtain final output results. Again, on the average, the time spent on the terminal decreases from 20 minutes to 10 minutes per customer, that is, six customers per hour are now served. Therefore, given $\lambda = 3.75$ and $\mu = 6$, facility utilization becomes 62.5%. The administration now believes that, since utilization of the terminal has decreased, any student complaints are indeed unjustified. Do you agree?

Again, we apply the appropriate equations to our M/M/1 queueing system and discover the following probabilities:

$$P_0 = 0.3750$$
$$P_1 = 0.2344$$
$$P_2 = 0.1465$$
$$P_3 = 0.0916$$
$$P_4 = 0.0572$$
$$P_5 = 0.0358$$

The important queueing system answers for this new M/M/1 system are below:

$$L \quad = 1.6667$$
$$L_q = 1.0417$$
$$W \quad = 0.4445$$
$$W_q = 0.2778$$

There are now, on the average, one and two-thirds students waiting in the system for service as opposed to the previous five students. The time a typical student spends in the system has decreased from the previous 120 minutes to about 27 minutes. Service requirements in the new queueing system are 10 minutes so a wait of 27 minutes may not seem as long as waiting 120 minutes for 20 minutes of service. Service requirements have been halved, while the time spent waiting for that service has been approximately quartered. Would student complaints concerning long waiting lines be justified now? Would the administration be concerned about a decrease in utilization between the two systems? Would another terminal (or more) be justified in order to change substantially the output parameters of this problem? This last question will be answered when we complete our examination of the M/M/c queueing system in the next section.

The M/M/c Queueing System

Our previous examination involved the single-server queueing system which could be mathematically modeled by a Poisson input process and an exponentially distributed service process. The M/M/c queueing system is a waiting-line system which, again, may be mathematically modeled by a Poisson input process and an exponentially distributed service process. However, the number of available servers (c) in the service facility are not limited to a single server. The M/M/c queueing system allows us to examine the situation where the number of servers in the service facility may assume any finite value greater than one. In particular, the assumptions below are representative of the primary characteristics of the M/M/c queueing system.

The input process will be given a Poisson probability distribution where customer arrivals will be assumed to follow the random manner (constant average rate) exhibited by the Poisson process. This average customer arrival rate will be denoted by the symbol λ. The dimension associated with this average customer arrival rate is in customers per unit of time t.

The service process will follow a negative exponential probability distribution where the service time will be symbolized by the constant average service rate $1/\mu$ customers per unit of time t.

The number of available servers in the service facility will be equal to the positive integer value denoted by the letter c (where $1 < c < \infty$). As mentioned previously, each of the c servers follow the identical negative exponential probability distribution given in the service process.

The source population of potential customers is assumed to be infinite in size. Allen[3] describes the finite source population condition with regard to an interactive computing system.

The allowable maximum queue length will be assumed to be infinite in size, that is, all possible customers generated from the infinite source population will be allowed into the system.

The queue discipline is of a first-in first-out (FIFO) nature.

Most real-world queueing systems fall into the M/M/c category. As an example, consider the typical grocery store checkout counter area. The checkout counters (c servers) serve a (seemingly) infinite source population through a single queue

3A.O. Allen, "Elements of Queueing Theory for System Design," *IBM Systems Journal* 14 (1975): 175–76.

(the staging area at the front of the store). Each of these c servers offers the same type of parallel service, namely, ringing up and bagging the items picked from the shelves of the store by the customers. Although the results which we shall obtain by examining the M/M/c queueing system are not exactly comparable to their real-world counterparts, an initial approximation may be made as to the general characteristics of these somewhat complex real-world systems.

The required input parameters to the M/M/c queueing system include (1) the average customer arrival rate (λ) of the customers generated from the infinite source population; (2) the average service rate (μ) of the c identical servers in the service facility; and (3) an additional parameter denoted by the letter c which corresponds to the number of servers in the service facility.

Just as we wanted to know the necessary equations which would determine the average queue length and the average number of customers in the queueing system for the M/M/1 queueing system, the necessary equations to solve for these same averages in the M/M/c queueing system are also required. As we shall soon see, the final equations for the M/M/c system are more complex than were the corresponding ones for the simpler M/M/1 queueing system.

Our initial equation involves the probability of zero customers in the system at time t. This equation is as follows:

$$P_0(t) = \cfrac{1}{\displaystyle\sum_{n=0}^{c-1} \cfrac{\cfrac{\lambda^n}{\mu}}{n!} + \cfrac{\cfrac{\lambda^c}{\mu}}{c!(1 - (\rho/c))}} \tag{5-13}$$

The parameters here are: c equals the number of servers in the service facility, λ represents the average customer arrival rate, μ equals the average service rate of the c identical servers, and ρ is the traffic intensity ratio (λ/μ).

The probability of n customers being in the system at time t must be regarded in terms of two equations. The first involves the particular case where the number of customers (n) is less than the number of servers (c). The second equation takes into account the case where the number of customers (n) is greater than or equal to the number of servers (c).

The equation governing the situation when the number of customers (n) is less than the number of servers (c) in the system is

$$P_n(t) = (1/n!)(\rho)^n P_0(t) \tag{5-14}$$

where n equals the number of customers in the queueing system at time t, ρ is the traffic intensity ratio (λ/μ), and $P_0(t)$ is the probability that zero customers are in the system at time t. This last parameter was calculated in equation 5-13.

The equation for the situation where the number of customers (n) is greater than or equal to the number of servers (c) is given by

$$P_n(t) = \frac{1}{c!c^{n-c}} \, \rho^n P_0(t) \tag{5-15}$$

where c is the number of identical servers in the service facility, n is the number of customers in the system at time t (and $n \geqslant c$), ρ is the traffic intensity ratio (λ/μ) and $P_0(t)$ is the probability that zero customers are in the system at time t (see equation 5-13).

The equations which allow us to calculate the average waiting time in both the queue and the system are given as

$$W_q = \frac{\dfrac{\rho^c P_0(t)}{c!(1 - (\rho/c))}}{c\mu(1 - (\rho/c))} \tag{5-16}$$

$$W = W_q + 1/\mu \tag{5-17}$$

where c is the number of identical servers in the service facility, λ is the average customer arrival rate, μ is the average service rate, ρ is the traffic intensity ratio (λ/μ), and $P_0(t)$ is the probability of zero customers in the system at time t (equation 5-13).

The equations which solve for the queue length and for the average number of customers in the system are as follows:

$$L_q = \lambda W_q \tag{5-18}$$

$$L = \lambda W \tag{5-19}$$

Although the M/M/c queueing system is the simplest of the multiserver queueing systems, there are a number of assumptions which must be made before it can be mathematically valid:

1. If all the servers are busy when a customer arrives, the arriving customer must join the single queue from which all available servers are fed.
2. If one server is idle while all the other servers are busy at the time a customer arrives, this idle server must service this newly arrived customer immediately.
3. If there are two or more idle servers, the customer may choose any one of them for immediate service. No preference is given to any available server during this particular option, since all available servers offer the same type of service to all arriving customers.

Since the queue discipline is assumed to be FIFO, as soon as a server becomes free (that is, idle), the customer who has waited the longest will be the next one to receive service from this idle server. Of course, if there are no waiting customers, upon arrival of a customer one server bocomes busy immediately. No server is allowed to be idle if there are customers waiting in the queue.

Figure 5-3 is the programmed routine which will greatly simplify our calculations regarding the M/M/c queueing system. Figure 5-4 is the associated output from the programmed routine of Figure 5-3 given the various input parameters as shown in the listing.

```
1000 REM   M/M/C QUEUEING SYSTEM
1010 REM   VARIABLE-NAME    DESCRIPTION
1020 REM      C         NUMBER OF SERVERS
1030 REM      L         AVERAGE NUMBER OF CUSTOMERS IN SYSTEM
1040 REM      LM        LAMBDA-AVERAGE CUSTOMER ARRIVAL RATE
1050 REM      LQ        AVERAGE NUMBER OF CUSTOMERS IN QUEUE
1060 REM      MU        MU-AVERAGE SERVICE RATE
1070 REM      N         CUSTOMER COUNTER
1080 REM      NC        N FACTORIAL COUNTER
1090 REM      NF        N!
1100 REM      NM        N FACTORIAL COUNTER
1110 REM      NS        PROBABILITY COUNTER
1120 REM      P(0)      PROBABILITY OF 0 CUSTOMERS IN SYSTEM
1130 REM      P(N)      PROBABILITY OF N CUSTOMERS IN SYSTEM
1140 REM      PS        PROBABILITY ACCUMULATOR
1150 REM      RO        RHO-TRAFFIC INTENSITY RATIO
1160 REM      W         AVERAGE WAITING-TIME IN SYSTEM
1170 REM      WQ        AVERAGE WAITING-TIME IN QUEUE
1180 DIM P(5)
1190 PRINT"ENTER AVERAGE CUSTOMER ARRIVAL RATE (LAMBDA)";
1200 INPUT LM
1210 PRINT"ENTER AVERAGE SERVICE RATE (MU)";
1220 INPUT MU
1230 PRINT"ENTER THE NUMBER OF SERVERS (C)";
1240 INPUT C
1250 PN = 0.0
1260 P(0) = 0.0
1270 NM = C
1280 GOSUB 1620
1290 PS = ((LM/MU)↑C/NF)*(1.0/(1.0-((LM/MU)/C)))}
1300 FOR NS = 1 TO C-1
1310    NM = NS
1320    GOSUB 1620
```

FIGURE 5-3: Programmed routine for the M/M/c queueing system

```
1330   PN = (1.0/NF)*((LM/MU)↑NS)+PN
1340  NEXT NS
1350  P(0) = 1.0 / (PN + PS)
1360  FOR NS = 1 TO 5
1370    IF NS < C THEN 1420
1380    NM = C
1390    GOSUB 1620
1400    P(NS) = 1.0/(NF*C↑(NS-C))*(LM/MU)↑NS*P(0)
1410    GOTO 1450
1420    NM = NS
1430    GOSUB 1620
1440    P(NS) = (1.0/NF)*(LM/MU)↑NS*P(0)
1450  NEXT NS
1460  NM = C
1470  GOSUB 1620
1480  WQ = (((LM/MU)↑C*P(0))/(NF*(1.0-((LM/MU)/C))))/(C*MU*
         (1.0-((LM/MU)/C)))
1490  W = WQ + 1.0 / MU
1500  LQ = LM * WQ
1510  L = LM * W
1520  PRINT"PROBABILITY (P) OF (N) CUSTOMERS IN THE SYSTEM"
1530  PRINT"(N)          (P)"
1540  FOR N = 0 TO 5
1550    PRINT N, P(N)
1560  NEXT N
1570  PRINT"AVERAGE NUMBER OF CUSTOMERS IN THE SYSTEM ="; L
1580  PRINT"AVERAGE NUMBER OF CUSTOMER IN THE QUEUE ="; LQ
1590  PRINT"AVERAGE WAITING-TIME IN THE SYSTEM ="; W
1600  PRINT"AVERAGE WAITING-TIME IN THE QUEUE ="; WQ
1610  GOTO 1190
1620  REM SUBROUTINE TO GENERATE N!
1630  NF = 1
1640  FOR NC = NM TO 1 STEP -1
1650    NF = NF * NC
1660  NEXT NC
1670  RETURN
1680  END
```

FIGURE 5-3: (Continued)

ENTER AVERAGE CUSTOMER ARRIVAL RATE (LAMBDA)? 17
ENTER AVERAGE SERVICE RATE (MU)? 6
ENTER THE NUMBER OF SERVERS (C)? 3
PROBABILITY (P) OF (N) CUSTOMERS IN THE SYSTEM

(N)	(P)
0	.0133
1	.0377
2	.0535
3	.0505
4	.0477
5	.0450

AVERAGE NUMBER OF CUSTOMERS IN THE SYSTEM = 18.2830
AVERAGE NUMBER OF CUSTOMERS IN THE QUEUE = 15.4497
AVERAGE WAITING–TIME IN THE SYSTEM = 1.0755
AVERAGE WAITING–TIME IN THE QUEUE = .9088

FIGURE 5-4: Output from Figure 5-3

Remember the student terminal problem we examined during our discussion of the M/M/1 queueing system? Initially, the administration of a university computing center offered an on-line terminal to students for their use during an eight-hour class day. The necessary "answers" for the single-terminal situation were given previously. Due to the complaints of the students, a study was initiated to determine the effectiveness of adding another terminal to the existing system. Rather than purchase and hardwire the terminal into the system, the administration has decided to use the equations and programmed routine given for the M/M/c queueing system to discover any changes in student behavior regarding waiting times.

Under normal conditions, students arrive at the terminal according to a Poisson distribution with an average value equal to 2.5 per hour (that is, $\lambda = 2.5$). Service requirements under these normal conditions average three customers per hour. The administration would like to know the queueing parameters when two terminals are used ($c = 2$). These values are given as:

$$L = 0.1518$$
$$L_q = 0.5685$$
$$W = 0.0607$$
$$W_q = 0.3333$$

Server utilization for the dual-terminal system is equal to 41.7%, that is, each of the two terminals will be in use for three hours and 20 minutes during a typical eight-hour day. Compare these new values (dual terminals) with the old values (single terminal):

	L	L_q	W	W_q
Single terminal	5.0000	4.1667	2.0000	1.6667
Dual terminals	0.1518	0.5685	0.0607	0.3333

Obviously, student waiting times will decrease dramatically under the dual-terminal system as can readily be seen. However, utilization of the service facility has not changed; two terminals busy for 41.7% of the eight-hour period is equal to a single terminal being busy for 83.3% of the same period. Therefore, student complaints should fall to zero, yet overall utilization of the entire system may not be sufficient to justify the cost of the additional terminal.

Under the old single-terminal system, a student was made to wait (on the average) for 100 minutes in order to receive 20 minutes of service, while the new system required a waiting time of 20 minutes for the same 20-minute service requirement.

The input parameters for the final days of the semester period ($\lambda = 3.75$ and $\mu = 6$) yield the following output parameters:

$$L = 0.0293$$
$$L_q = 0.3418$$
$$W = 0.0078$$
$$W_q = 0.1667$$

Under the dual-terminal system, a student will wait in line for approximately 10 minutes in order to be offered service for 10 minutes. The single-terminal system made the student wait for approximately 17 minutes in order to gain service for 10 minutes.

The question now posed is, is a dual-terminal system justified in light of the "answers" provided by the queueing models? Under normal conditions, student waiting times are decreased by 80% (100 minutes versus 20 minutes), while under "abnormal" conditions the waiting time is decreased by approximately 40% (17 minutes versus 10 minutes). Since we have only waiting-time parameters, it would seem that a dual-terminal system would be justified, at least from the students' point of view. The administration, however, must consider the costs involved with the second terminal. This latter problem will be left to the reader to ponder.

A natural extension to our dual-terminal queueing system involves the condition when student activity increases due to an increase in student enrollment or simply because more students want to use the terminal. We shall now examine the situation where student activity increases to an average arrival rate equal to 12 students per hour (that is, λ = 12). Since the average customer arrival rate must, by definition, be less than the average service rate, the computing center administration must decrease the allowable service time requirements for its users. Let us assume that the new value for average service time is equal to three minutes per customer; in other words, the system may handle 20 arrivals per hour. If this decrease in average service rate were not put into effect, the resulting waiting line would grow without bound and, surely, student complaints would be justified simply by viewing the physical waiting line. Our input parameters, therefore, become λ = 12, μ = 20, and c = 2. The resulting queueing values are calculated to be as follows:

$$L = 0.0236$$
$$L_q = 0.3236$$
$$W = 0.0020$$
$$W_q = 0.0500$$

The most obvious questions now become, would a third terminal be required in order to decrease student waiting time? Would a third terminal, in fact, be necessary at all considering the calculated queueing values given previously? What values of average customer arrival and average service rate would have to be in effect before the students begin experiencing substantial delays in the waiting line? Underlying all these questions is, of course, the cost of a third terminal and its justification.

Example Problems

A particularly realistic M/M/c queueing system involves the operation of a multi-pump gasoline station. For our purposes, we shall assume that the service facility consists of three gasoline pumps, that is, c = 3. Customers arrive at the queueing system at an average rate (λ) equal to four customers per hour. The average service rate (μ) will be considered to be equal to five service completions per hour. Using these input parameters and the programmed routine of Figure 5-3, we may answer the following questions concerning this queueing system:

1. What is the probability of finding 0, 1, . . . , 5 customers in the system?
2. What is the average length of the waiting line?
3. What is the average number of customers in the system?
4. How long, on the average, must a typical customer wait in the queue prior to being selected for service by one of the three servers?
5. What is the average waiting time in the system?
6. What effect does a change in the input parameters have upon the results of the system?

The answers to the question of finding n customers in the system (where $n = 0,$ $1, \ldots , 5$) follow:

$$P_0 = 0.8088$$
$$P_1 = 0.6471$$
$$P_2 = 0.2588$$
$$P_3 = 0.0690$$
$$P_4 = 0.0184$$
$$P_5 = 0.0049$$

These results were obtained by applying equations 5-13 through 5-15. Applying these equations manually should result in similar results, however it is much easier to have the computer do this work.

Our programmed routine for the M/M/c queueing system also gives the following queueing system output parameters:

$$L = 0.8342$$
$$L_q = 0.0342$$
$$W = 0.2086$$
$$W_q = 0.0086$$

These results were obtained from equations 5-16 through 5-19.

Since we are most interested in studying the results given for this queueing system, let us begin by examining the results for the parameters L, L_q, W, and W_q. The value for W is given as 0.2086, which effectively means that a customer will have to wait in the system (line plus service) for approximately 13 minutes. Since the average service rate (μ) is five customers per hour, this waiting time consists of the time spent in the queue and the time spent in the service facility. Since service is completed in 12 minutes, the remaining time must be spent in the waiting line. W_q is equal to 0.0086, which is approximately 30 seconds. In order for the customer to receive 12 minutes of service, that customer must wait in line for approximately 30 seconds. On the average, there will be 0.8342 customers in the system with 0.0342 customers in the queue. This is not a very long waiting line.

What happens, however, when a gas shortage occurs? Let us assume that, under such a condition, we now see an increase in the number of arriving customers from its initial (preshortage) value of four customers per hour to a new value of eight customers per hour. This doubling (or tripling) of arriving customers is not unrealistic as can be proven by the lines which formed during those days of gas shortages. The average service requirements per customer during our gas-shortage situation will increase from its initial value of five customers per hour to seven customers per hour. This particular condition will exist because each customer will require less time for service. Each customer is now receiving less service time since each is buying less gasoline. Remember the $5 or $8 limits placed on gasoline purchases during those gas-shortage days? The solutions to the various output parameters for this queueing system follow:

$$P_0 = 0.4550$$
$$P_1 = 0.5200$$
$$P_2 = 0.2971$$
$$P_3 = 0.1132$$
$$P_4 = 0.0431$$
$$P_5 = 0.0164$$
$$L = 1.2554$$
$$L_q = 0.1125$$
$$W = 0.1569$$
$$W_q = 0.0141$$

Our new queueing system has shown a decrease in total customer waiting time from the preshortage value of 13 minutes to a new value of approximately nine minutes. This is because service requirements have been cut from the preshortage value of five customers per hour to the new value of eight customers per hour. However, the time spent waiting for service has increased from the preshortage value of 30 seconds to a new value of (approximately) 50 seconds. This condition is due to the fact that we now have more customers arriving at the queueing system (an increase from four customers per hour in the preshortage situation to eight customers per hour in the shortage case). We may extend this increase in customer arrivals to a value equal to 12 customers per hour in order to obtain the following output parameters:

$$P_0 = 0.1944$$
$$P_1 = 0.3333$$
$$P_2 = 0.2857$$
$$P_3 = 0.1633$$
$$P_4 = 0.0933$$
$$P_5 = 0.0533$$
$$L = 2.2222$$
$$L_q = 0.5079$$
$$W = 0.1852$$
$$W_q = 0.0423$$

A tripling of customer arrivals from the preshortage value increases the customer waiting time in the queue from 30 seconds to approximately three minutes. The total time a customer spends in the system is now equal to approximately 11 minutes (versus 13 minutes for the preshortage case). It can be seen that the major portion of time is spent waiting in line.

What effect will be present when the service rate (μ) is retained at the preshortage level of five customers per hour? Given the input parameters

$$\mu = 5$$
$$\lambda = 8$$
$$c = 3$$

and

$$\mu = 5$$
$$\lambda = 12$$
$$c = 3$$

what values of output parameters will be present for the queueing system? This problem will be left for the reader to determine the appropriate values.

A somewhat valid solution to the problem of long queues with regard to the gas-shortage problem involved the even/odd allocation of sales; customers could purchase gasoline depending upon the last digit of their license tags—even numbers on one day, odd numbers on another day. Given the input parameters

$$\mu = 5$$
$$\lambda = 4$$
$$c = 3$$

and

$$\mu = 5$$
$$\lambda = 6$$
$$c = 3$$

what are the appropriate output parameters? The solutions are below:

	$\lambda = 4$	$\lambda = 6$
P_0	0.8088	0.4167
P_1	0.6471	0.5000
P_2	0.2588	0.3000
P_3	0.0690	0.1200
P_4	0.0184	0.0480
P_5	0.0049	0.0192
L	0.8342	1.3333
L_q	0.0342	0.1333
W	0.2086	0.2222
W_q	0.0086	0.0222

What questions may be raised concerning the implementation of an even/odd allocation system? Is there a better method? What can we expect of the output parameters if the service rate is decreased (or increased) from its value of five customers per hour? What other questions may be asked regarding this real-world queueing system?

The last problem involving our gasoline station queueing system will ask the question, how does an increase or a decrease in the number of servers affect the parameters of the problem? In other words, under normal operating conditions, the number of arrivals expect a certain amount of available servers (open gasoline pumps, for example). May the number of available servers be increased or decreased in order to take advantage of a possible decrease in overall operating costs? The problem is to find an acceptable level of customer waiting times after which customer satisfaction suffers and attendant profits decrease. Likewise, increasing (if possible) the number of servers during a gas-shortage condition may be examined in order to determine the point of acceptable customer waiting times as well as the attendant operating costs versus potential profits. A number of various input parameters are presented below for an initial examination of this particular queueing system. The reader is, of course, urged to utilize the programmed routine with other input parameters for a further study of this queueing system. The values given for the customer arrival rate and service rate are $\lambda = 8$ and $\mu = 4$.

	$c = 4$	$c = 5$	$c = 6$
L	2.2000	2.0460	2.0104
L_q	0.2000	0.0460	0.0104
W	0.2750	0.2557	0.2513
W_q	0.0250	0.0057	0.0013

Our final M/M/c queueing system example problem involves the situation where customers arrive at a queueing system and select a sequential number upon entry to the waiting line. Service selection is, therefore, dependent upon the customer's number, that is, a first-come first-served queue discipline is present. The real-world analogy of this model may be seen at a catalog store or a deli counter where arriving customers select a numbered card, wait for service, and are eventually served by one of c identical servers. The somewhat serpentine waiting line which banks are now using is yet another real-world example. Customers arrive at the bank (according to a Poisson probability distribution with an average value equal to λ), form a single sequential queue, and are offered service from the next available teller. Service is rendered, of course, according to an exponential probability distribution having the parameter μ as its average value.

Rather than present a detailed description of this queueing system, the reader is urged to examine the following output parameters. The questions which naturally arise from this example are quite similar to those asked in the gasoline station queueing system:

1. What effect does an increase or a decrease in customer arrivals have on the output parameters of the system?
2. What effect does an increase or a decrease in service rates have on the output parameters of the system?
3. What effect does an increase or a decrease in the number of servers have on the output parameters of the system?

A few numerical examples of this queueing system follow:

	$\lambda = 15, \mu = 8, c = 5$	$\lambda = 30, \mu = 8, c = 5$	$\lambda = 30, \mu = 8, c = 8$
P_0	0.1800	0.0190	0.0240
P_1	0.3375	0.0714	0.0898
P_2	0.3164	0.1339	0.1684
P_3	0.1978	0.1673	0.2105
P_4	0.0927	0.1569	0.1974
P_5	0.0348	0.1176	0.1480
L	1.9084	5.1617	3.7886
L_q	0.0334	1.4117	0.0386
W	0.1272	0.1721	0.1263
W_q	0.0022	0.0471	0.0013

The $M/M/\infty$ Queueing System

So far we have considered one queueing system (M/M/1) which offered but a solitary server and one (M/M/c) which offered some finite number of servers. In both instances, the number of servers did not equal the number of customers who could potentially be generated from our infinite source population. Therefore, our previous examination centered about the characteristics of the waiting times which arriving customers spent in the system waiting for a free (idle) server or servers to accommodate them.

A queueing system which has available enough servers to effectively service all possible arriving customers will be now considered from a purely introductory standpoint. In an M/M/∞ queueing system, each arriving customer is afforded service immediately upon entry into the system. There are as many servers ($c = \infty$) as there are potential customers, that is, an infinite amount. In addition, as is the case in a multiserver queueing system, each server offers identical service to any customer requesting service. Since any idle servers cannot be tolerated, a departing customer leaves a server who will become immediately available for the next arriving customer.

Since each arriving customer is served immediately, there cannot be any waiting line or waiting time in a queue. Thus, with no waiting line, the queueing system contains only customers in service. Our primary concern in an M/M/∞ queueing system will be considered simply in terms of the distribution of busy servers, the mean number of busy servers, and the variance of busy servers. Calculation of these values will effectively yield a limiting value as to the number of required servers for this particular queueing system.

It seems quite impractical, at first glance, that a real-world M/M/∞ queueing system could ever exist. Granted, it does represent the ideal situation for a queueing system, especially with regard to the arriving customers. The primary goal which may be achieved by examining this queueing system is in determining practical upper bounds on service facility numbers of more practical real-world systems. We could, for example, increase parameter c to such a large finite amount as to approximate the M/M/∞ queueing system in order to determine this upper bound.

The M/M/∞ queueing system has an input parameter which is a Poisson probability distribution with an average customer arrival rate equal to parameter λ. Each server exhibits a negative exponential service-time probability distribution with an average service rate equal to parameter μ. The queue discipline is assumed to be a first-in first-out (FIFO), while the source population of potential customers is also assumed to be infinite. The length of the waiting line, is, of course, not required.

The $M/E_k/1$ Queueing System

A particularly interesting queueing system involves a nonexponential service-time probability distribution named in honor of A.K. Erlang. The Poisson input process, Erlang service-time distribution, single-server queueing system will be the focus of our attention in this section. It is frequently referred to in the literature as the Erlangian distribution, the gamma distribution, the Pearson Type III distribution, or a variety of other names. For our purposes, we shall use the name Erlang service-time probability distribution.

Most real-world service-time probability distributions do not usually assume either the purely exponential characteristics or the purely constant characteristics of a probability distribution function. Rather, the actual service times fall somewhere in between these two extremes. The Erlang service-time probability distribution may, therefore, be used in order to model more effectively the service times of a real-world queueing system. This is not to say, however, that either the exponential or the constant service-time distributions should be forgotten; in fact, both should be considered in terms of the wide range which they allow the queueing problem solver. Initially, it is usual to apply both these probability distributions to the real-world service process in order to gain a better insight into the extreme values which a typical service process may assume. When these extreme values are calculated, then the Erlang probability distribution is applied to the service process in order to fine-tune the mathematics of the particular queueing system. Of course, when an Erlang service process is known to exist, then the application of the constant and exponential probability distributions to the given service process are omitted. The advantage of simplified mathematics with regard to the constant and the exponential probability distributions is well worth the effort in the study of a given queueing system. The lack of uncertainty when using the constant probability distribution as well as the memoryless property of the exponential probability distribution make for easier calculations when trying to determine the upper and the lower bounds of a particular service-time process.

The exponential probability distribution is a random process which may normally be used to define the natural or real-world state of a given queueing system service-time distribution. For example, a telephone exchange may be described adequately by assuming that the length of telephone calls closely approximates the exponential probability distribution. In other words, the service-time distribution (the length of a telephone call) can be described as an exponential probability distribution. Likewise, other real-world random processes tend to follow the exponential probability distribution if they are under observation for a long enough period of time.

At the other extreme, a service-time process may be defined adequately in terms of a constant probability distribution. If the exponential service-time probability distribution can be considered the worst case situation, then the constant probability distribution service time may be thought of as the best case situation (mathematically, that is).

Although many real-world queueing systems may be described in terms of the exponential or the constant service-time probability distributions, there are a large number of remaining queueing systems in the real world which cannot be truly approximated by either of these probability distributions. Since there does, in fact, exist a large enough number of queueing system situations between these two extremes, namely, the cases between $0 < \mu < \infty$, it would certainly be most convenient to classify these in-between queueing systems with a definite service-time probability distribution.

The Erlang-k probability distribution is the vehicle which will allow us to accommodate these not quite exponential and not quite constant probability distributions regarding the service-time requirements of a queueing system in the real world. A.K. Erlang, during his study of the traffic problems of the Danish telephone system in 1917, proposed a service-time probability distribution which did not follow the usually reliable exponential probability distribution. The Erlang-k probability distribution was the answer to this rather perplexing problem.

As with the constant service-time probability distribution for the multiserver queueing system, the multiserver Erlang-k service-time probability distribution becomes quite formidable a task to state in mathematical terms. In fact, the probability of expressing this queueing system ($M/E_k/c$) has, to date, been impossible. This is yet another fertile area of queueing theory open to the serious reader.

The general form of the probability density function for the Erlang-k probability distribution (single-server service facility) is given as

$$f(x) = \frac{(\mu k)^k}{(k-1)!} x^{k-1} \exp^{-kx\mu}$$

Where parameter x is greater than or equal to zero and less than infinity, and parameter k is an integer and is also greater than zero. The Erlang probability distribution has a probability density function which would be identical to the gamma probability distribution function were it not necessary for parameter k to be an integer.

When parameter k is equal to one, the probability density function of the Erlang-k probability distribution becomes the equation which defines the probability density function of the exponential probability distribution. When parameter k, on the other hand, is equal to infinity, the probability density function becomes the constant (degenerate) probability density function. For values of parameter k which range between one and infinity, the probability density function, $f(x)$, ranges between the exponential and the constant probability distributions, that is, all the possible values for the in-between situations. As parameter k increases in value toward infinity, the more the probability distribution begins to behave as a constant probability distribution. On the other hand, the smaller the parameter k, the more the distribution begins to look like the exponential probability distribution.

If we were to examine k independent random variables $(x_1, x_2, \ldots, x_k)$ where each of the random variables (x_i) has an identical exponential probability distribution with a mean value equal to $1/(k\mu)$, then the sum of these random variables

$$\sum_{i=1}^{k} x_i \quad \text{where } i = 1, 2, \ldots, k$$

will have an Erlangian probability distribution with parameters μ and k. This particular Erlang probability distribution involves not a single exponential service-time task but a sequence of k exponential service-time tasks. In other words, the single-server is required to service each customer k times, or, on the other hand, each customer is required to be served by a sequence of k identical servers in series.

The latter case is not to be confused with a purely serial service facility where as many customers are allowed into the service facility as there are available servers. Rather, only a single customer is allowed into the service facility at any particular time that service is offered by the facility. Only after the current customer is fully served (that is, this customer completes service at the kth service area) will the next customer be chosen for entry into the service facility.

Given the parameters k (number of stages of service) and μ (average service rate), the following equations for the mean and the standard deviation of the distribution may be calculated:

$$\text{mean} = 1/\mu$$
$$\text{standard deviations} = 1/\mu(k)^{1/2}$$

Under normal conditions, the particular value of k is not given (or known beforehand). Rather, the appropriate values for the mean and the standard deviation are the given parameters. Therefore, the corresponding Erlang parameter may be found from the values given for the mean and the standard deviation by taking the nearest integer value of the following:

$$k = (\text{mean})^2/(\text{variance})$$

For the single-server, Poisson arrival with average arrival rate equal to parameter λ, service times which follow the Erlang-k probability distribution with an average service rate equal to parameter μ, the variance will be equal to the following equation:

$$\sigma^2 = 1/k\mu^2$$

The following queueing relationships exist for the $M/E_k/1$ queueing system assuming an infinite source population, an infinite queue length, and a FIFO queue discipline:

$$L_q = \frac{k+1}{2k} \times \frac{\lambda^2}{\mu(\mu - \lambda)} \tag{5-20}$$

$$L = \frac{k+1}{2k} \times \frac{\lambda^2}{\mu(\mu - \lambda)} + \rho \tag{5-21}$$

$$W_q = \frac{k+1}{2k} \times \frac{\lambda}{\mu(\mu - \lambda)} \tag{5-22}$$

$$W = \frac{k+1}{2k} \times \frac{\lambda}{\mu(\mu - \lambda)} + (1/\mu) \tag{5-23}$$

where k is the Erlang stage parameter (as in Erlang-k) and an integer, μ is the average service rate, λ is the average customer arrival rate, and $\rho = \lambda/\mu$.

The programmed routine in Figure 5-5 and the sample outputs in Figure 5-6 reflect the information we have just examined for the $M/E_k/1$ queueing system.

Allen[4] and Chang[5] offer a number of illustrations dealing with the $M/E_k/1$ queueing system as it applies to particular problems in computer systems.

[4] A.O. Allen, "Elements of Probability for System Design," *IBM Systems Journal* 13(1974): 341–44.

[5] W. Chang, "Single Server Queueing Processes in Computing Systems," *IBM Systems Journal* 9(1970): 48–49.

```
1000 REM    M/E(K)/1 QUEUEING SYSTEM
1010 REM    VARIABLE-NAME    DESCRIPTION
1020 REM    K    ERLANG-K PARAMETER
1030 REM    L    AVERAGE NUMBER OF CUSTOMERS IN SYSTEM
1040 REM    LM   LAMBDA - AVERAGE CUSTOMER ARRIVAL RATE
1050 REM    LQ   AVERAGE NUMBER OF CUSTOMERS IN QUEUE
1060 REM    MU   MU - AVERAGE SERVICE RATE
1070 REM    RO   RHO - TRAFFIC INTENSITY RATIO
1080 REM    W    AVERAGE WAITING-TIME IN SYSTEM
1090 REM    WQ   AVERAGE WAITING-TIME IN QUEUE
1100 PRINT"ENTER AVERAGE CUSTOMER ARRIVAL RATE (LAMBDA)";
1110 INPUT LM
1120 PRINT"ENTER AVERAGE SERVICE RATE (MU)";
1130 INPUT MU
1140 IF LM/MU < 1.0 THEN 1190
1150   PRINT"TRAFFIC INTENSITY RATIO (LAMBDA/MU) MUST BE"
1160   PRINT"LESS THAN ONE"
1170   PRINT
1180   GOTO 1100
1190 RO = LM/MU
1200 PRINT"ENTER ERLANG-K PARAMETER (K)";
1210 INPUT K
1220 LQ = ((K+1)/2*K))*((LM↑2)/(MU*(MU-LM)))
1230 L = LQ + RO
1240 WQ = ((K+1)/2*K))*(LM/(MU*(MU-LM)))
1250 W = WQ + (1.0/MU)
1260 PRINT"AVERAGE NUMBER OF CUSTOMERS IN THE SYSTEM="; L
1270 PRINT"AVERAGE NUMBER OF CUSTOMERS IN THE QUEUE ="; LQ
1280 PRINT"AVERAGE WAITING-TIME IN THE SYSTEM ="; W
1290 PRINT"AVERAGE WAITING-TIME IN THE QUEUE ="; WQ
1300 GOTO 1100
1310 END
```

FIGURE 5-5: Programmed routine for the $M/E_k/1$ queueing system

```
ENTER AVERAGE CUSTOMER ARRIVAL RATE (LAMBDA)? 5
ENTER AVERAGE SERVICE RATE (MU)? 10
ENTER ERLANG-K PARAMETER (K)? 2
AVERAGE NUMBER OF CUSTOMERS IN THE SYSTEM = 2
AVERAGE NUMBER OF CUSTOMERS IN THE QUEUE = 1.5
AVERAGE WAITING-TIME IN THE SYSTEM = .4
AVERAGE WAITING-TIME IN THE QUEUE = .3
```

FIGURE 5-6: Output from Figure 5-5

Example Problems

Our example problem of the $M/E_k/1$ queueing system will involve the characteristics displayed by a series of k traffic lights and their behavior in terms of queueing systems. In order to retain a degree of simplicity, we shall initially assume that
our traffic lights number two. The second part of this problem will use three traffic
lights ($k = 3$) in order to compare the queueing characteristics.

Let us initially set our customer arrival rate (λ) equal to three customers per
minute. Our initial service rate (μ) will be, for illustrative purposes, equal to seven
service completions per minute. In other words, the traffic lights will turn green
for a period of time which will allow seven vehicles to pass through the system
over a one-minute period.

Since we will be most interested in determining the output parameters L, L_q,
W, and W_q, these shall be the ones given to us by the programmed routine of Figure
5-5. Equations 5-20 through 5-23 will be used for the calculations of these output
parameters.

The output parameters for this queueing system are as follows:

$$L \;\; = 1.3929$$
$$L_q = 0.9643$$
$$W \;\; = 0.4643$$
$$W_q = 0.3214$$

A customer will wait in the system of two traffic lights for a period of (approximately) 28 seconds.

If the customer arrival rate (λ) is doubled to six arrivals per minute (on the
average), with $\mu = 7$ and $k = 2$, the following output parameters will be calculated:

$$L \;\; = 16.2857$$
$$L_q = 15.4286$$
$$W \;\; = \;\; 2.7143$$
$$W_q = \;\; 2.5714$$

Our customers will now have to wait in the system for approximately two minutes
and 34 seconds; a doubling of customer arrivals (three versus six) has resulted in
a waiting-time increase by a factor of about five. The following output parameters
take into account the progressive increase in customer arrivals for values of four
and five ($\mu = 7$ and $k = 2$):

	$\lambda = 4$	$\lambda = 5$
L	2.8571	6.0714
L_q	2.2857	5.3572
W	0.7143	1.2143
W_q	0.5714	1.0714

The reader is again urged to examine these figures as well as figures which result from other input values in order to appreciate the effects which various input values have on the overall queueing characteristics of this particular system.

Our $M/E_k/1$ example problem now takes on an additional traffic light, i.e., $k = 3$. Let us now briefly examine the resulting output parameters given the identical input parameters from the previous situation. These values are listed below for $\mu = 7$ and $k = 3$:

	$\lambda = 3$	$\lambda = 4$	$\lambda = 5$	$\lambda = 6$
L	2.3571	5.1429	11.4286	31.7143
L_q	1.9286	4.5714	10.7143	30.8571
W	0.7857	1.2857	2.2857	5.2857
W_q	0.6429	1.1429	2.1429	5.1429

A simple addition of one more stage (k) into our $M/E_k/1$ queueing system has resulted in a substantial increase in customer waiting times (to say the least!). For example, our customer must now wait for approximately 47 seconds in order to clear all the traffic lights when the average customer arrival rate is equal to three customers per minute. When k was equal to two, our customer had a total system waiting time equal to 28 seconds. Output values for the other input parameters should be examined, also. The additional traffic light has most certainly placed a substantial burden on customers with regard to the time they must spend waiting in the system. A natural question now follows: What effect will another traffic light have on the total system waiting time?

Since we may expect customer arrivals to increase in the future (that is, $\lambda = 7$, $8, \ldots$), what must be done to the average service rate (μ) in order to prevent a congestive queue from forming? At the present time, the average service rate is equal to seven service completions per minute. What value would you suggest for this input parameter in order to decrease customer waiting time to a more acceptable level?

Problems for Reader Solution

5-1. Use equation 5-7 to calculate the value for P_0 given the following input parameters:
   a) $\mu = 10$ and $\lambda = 2$
   b) $\mu = 10$ and $\lambda = 4$
   c) $\mu = 10$ and $\lambda = 8$
   d) $\mu = 10$ and $\lambda = 9$

5-2. Use equation 5-8 to calculate the value for P_n (where $n = 1, 2, 3,$ and 4) given the following input parameters:
   a) $\mu = 10$ and $\lambda = 2$
   b) $\mu = 10$ and $\lambda = 4$
   c) $\mu = 10$ and $\lambda = 8$
   d) $\mu = 10$ and $\lambda = 9$

5-3. Use equations 5-9 through 5-12 to calculate the values for L, L_q, W, and W_q given the following input parameters:
   a) $\mu = 10$ and $\lambda = 2$
   b) $\mu = 10$ and $\lambda = 4$
   c) $\mu = 10$ and $\lambda = 8$
   d) $\mu = 10$ and $\lambda = 9$

5-4. Use equation 5-13 to calculate the value for P_0 given the following input parameters:
   a) $c = 2, \mu = 10$, and $\lambda = 8$
   b) $c = 2, \mu = 10$, and $\lambda = 9$

5-5. Use equation 5-14 to calculate the value for P_n (where $n = 1, 2$) given the following input parameters:
   a) $c = 2, \mu = 10$, and $\lambda = 8$
   b) $c = 2, \mu = 10$, and $\lambda = 9$

5-6. Use equation 5-15 to calculate the value for P_n (where $n = 3, 4, 5,$) given the following input parameters:
   a) $c = 2, \mu = 10$, and $\lambda = 9$
   b) $c = 2, \mu = 10$, and $\lambda = 8$

5-7. Use equations 5-16 through 5-17 to calculate the values for L, L_q, W, and W_q given the following input parameters:
   a) $c = 2, \mu = 10$, and $\lambda = 8$
   b) $c = 2, \mu = 10$, and $\lambda = 9$

5-8. Use equations 5-20 through 5-23 to calculate the values for L, L_q, W, and W_q given the following input parameters (in all cases, $k = 1, 2$, and 3):
   a) $\mu = 10$ and $\lambda = 2$
   b) $\mu = 10$ and $\lambda = 4$
   c) $\mu = 10$ and $\lambda = 8$
   d) $\mu = 10$ and $\lambda = 9$

Answers to Problems

5-1.
   a) 0.800
   b) 0.600
   c) 0.200
   d) 0.100

5-2.
   a) $P_1 = 0.160, P_2 = 0.032, P_3 = 0.006, P_4 = 0.001$
   b) $P_1 = 0.600, P_2 = 0.240, P_3 = 0.096, P_4 = 0.038$
   c) $P_1 = 0.160, P_2 = 0.128, P_3 = 0.102, P_4 = 0.082$
   d) $P_1 = 0.090, P_2 = 0.081, P_3 = 0.073, P_4 = 0.066$

5-3.

   a) $L = 0.250, L_q = 0.050, W = 0.125, W_q = 0.025$
   b) $L = 0.667, L_q = 0.267, W = 0.167, W_q = 0.067$
   c) $L = 4.000, L_q = 3.200, W = 0.500, W_q = 0.400$
   d) $L = 9.000, L_q = 8.100, W = 1.000, W_q = 0.900$

5-4.

   a) 0.611
   b) 0.750

5-5.

   a) $P_1 = 0.600, P_2 = 0.240$
   b) $P_1 = 0.550, P_2 = 0.248$

5-6.

   a) $P_3 = 0.096, P_4 = 0.038, P_5 = 0.015$
   b) $P_3 = 0.111, P_4 = 0.050, P_5 = 0.022$

5-7.

   a) $L = 1.067, L_q = 0.267, W = 0.133, W_q = 0.033$
   b) $L = 1.268, L_q = 0.368, W = 0.141, W_q = 0.041$

5-8.

   a) $k = 1, L = 1.250, L_q = 0.050, W = 0.125, W_q = 0.025$
      $k = 2, L = 0.350, L_q = 0.150, W = 0.175, W_q = 0.075$
      $k = 3, L = 0.500, L_q = 0.300, W = 0.250, W_q = 0.150$
   b) $k = 1, L = 0.667, L_q = 0.267, W = 0.167, W_q = 0.067$
      $k = 2, L = 1.200, L_q = 0.800, W = 0.300, W_q = 0.200$
      $k = 3, L = 2.000, L_q = 1.600, W = 0.500, W_q = 0.400$
   c) $k = 1, L = 4.000, L_q = 3.200, W = 0.500, W_q = 0.400$
      $k = 2, L = 10.400, L_q = 9.600, W = 1.300, W_q = 1.200$
      $k = 3, L = 20.000, L_q = 19.200, W = 2.500, W_q = 2.400$
   d) $k = 1, L = 9.000, L_q = 8.100, W = 1.000, W_q = 0.900$
      $k = 2, L = 25.200, L_q = 24.300, W = 2.800, W_q = 2.700$
      $k = 3, L = 49.500, L_q = 48.600, W = 5.500, W_q = 5.400$

6

Queueing Systems with Nonexponential Distributions

The queueing systems examined in chapter 5 involved interarrival times and service times which were modeled by an exponential probability distribution, that is, the models were based on a birth-death process. This assumption is invalid in many real-world queueing systems because many do not readily fall into the birth-death category with regard to queueing distributions. When customer arrivals are scheduled or otherwise deterministic rather than probabilitstic, a Poisson input process which assumes exponential interarrival times does not apply. When the service-time distribution is not of an exponential nature, the assumption we made earlier does not apply. In cases where the exponential assumptions are not valid, we must be able to construct mathematically an appropriate queueing model which follows or approximates the probability distribution given for these special cases. An unfortunate fact concerning these queueing systems is that the attendant mathematics are quite complex. We shall not, therefore, analyze these queueing systems; rather, general descriptions and sufficient summary information will be provided in order that we may be able to handle a few of these special queueing systems.

The M/G/1 Queueing System

A queueing system which has as its input process a Poisson probability distribution (with an average customer arrival rate equal to the parameter λ) and a single server will be examined now. The service-time distribution which describes the M/G/1 queueing system is that of a general distribution.

Since any distribution may be assumed for the service-time distribution, it is necessary for the mean value $(1/\mu)$ as well as the variance (σ^2) of the distribution to be known. Given these two input parameters and the fact that

$$\rho = \lambda/\mu < 1$$

any queueing system can reach a steady-state condition where the following equations will hold true:

$$P_0(t) = 1 - \rho \qquad\qquad\qquad (6-1)$$

$$L_q = \frac{\lambda^2\sigma^2 + \rho^2}{2(1 - \rho)} \qquad\qquad (6-2)$$

$$L = \rho + L_q \qquad\qquad\qquad (6-3)$$

$$W_q = L_q/\lambda \qquad\qquad\qquad (6-4)$$

$$W = W_q + (1/\mu) = L/\lambda \qquad (6-5)$$

Allen[1] provides the resulting equations and the necessary summary information for the M/G/1 queueing system.

The equation which solves for the probability of the number of customers in the system and the waiting time of the customers in the queue as well as in the system requires the use of rather complex mathematics. In particular, it is necessary to use the inverted Laplace–Stieltjes transform for the service-time probability distribution. Clarke[2] provides a thorough mathematical treatment of the M/G/1 queueing system by presenting this non-Markovian queueing system in terms of the embedded chain concept. Equations 6–1 through 6–5 will be used in a programmed routine which will solve for the necessary queueing system measurements. Figure 6-1 is the programmed routine, while Figure 6-2 is an example of typical output from this routine.

```
1000 REM    M/G/1 QUEUEING SYSTEM
1010 REM    VARIABLE–NAME    DESCRIPTION
1020 REM    L    AVERAGE NUMBER OF CUSTOMERS IN SYSTEM
1030 REM    LM   LAMBDA – AVERAGE CUSTOMER ARRIVAL RATE
1040 REM    LQ   AVERAGE NUMBER OF CUSTOMERS IN QUEUE
1050 REM    MU   MU – AVERAGE SERVICE RATE
1060 REM    PO   PROBABILITY OF 0 CUSTOMERS IN SYSTEM
```

FIGURE 6-1: Programmed routine for the M/G/1 queueing system

[1] A.O. Allen, "Elements of Queueing Theory for System Design," *IBM Systems Journal* 14 (1975): 172–75.

[2] A.B. Clarke and R.L. Disney, *Probability and Random Processes for Engineers and Scientists* (New York: Wiley, 1970), pp. 317–24.

```
1070 REM     RO   RHO - TRAFFIC INTENSITY RATIO
1080 REM     V2   VARIANCE OF SERVICE DISTRIBUTION
1090 REM     W    AVERAGE WAITING-TIME IN SYSTEM
1100 REM     WQ   AVERAGE WAITING-TIME IN QUEUE
1110 PRINT"ENTER AVERAGE CUSTOMER ARRIVAL RATE (LAMBDA)";
1120 INPUT LM
1130 PRINT"ENTER AVERAGE SERVICE RATE (MU)";
1140 INPUT MU
1150 IF LM/MU < 1.0 THEN 1220
1160   PRINT"TRAFFIC INTENSITY RATIO (LAMBDA/MU) MUST BE"
1170   PRINT"LESS THAN ONE IF THE PROBABILITY OF 0 CUSTOMERS"
1180   PRINT"IN THE SYSTEM AT TIME T IS TO BE INDEPENDENT"
1190   PRINT"OF THE TIME T."
1200 PRINT
1210 GOTO 1110
1220 PRINT"ENTER VARIANCE OF DISTRIBUTION";
1230 INPUT V2
1240 RO = LM/MU
1250 PO = 1.0 - RO
1260 LQ = (((LM↑2)*V2) + (RO↑2))/(2*(1.0 - RO))
1270 L = RO + LQ
1280 W = LQ/LM
1290 WQ = L/LM
1300 PRINT"PROBABILITY OF ZERO CUSTOMERS IN THE SYSTEM =";PO
1310 PRINT"AVERAGE NUMBER OF CUSTOMERS IN THE SYSTEM =";L
1320 PRINT"AVERAGE NUMBER OF CUSTOMERS IN THE QUEUE =";LQ
1330 PRINT"AVERAGE WAITING-TIME IN THE SYSTEM =";W
1340 PRINT"AVERAGE WAITING-TIME IN THE QUEUE =";WQ
1350 GOTO 1110
1360 END
```

FIGURE 6-1: (Continued)

ENTER AVERAGE CUSTOMER ARRIVAL RATE (LAMBDA)? 5
ENTER AVERAGE SERVICE RATE (MU)? 10
ENTER VARIANCE OF DISTRIBUTION? 3
PROBABILITY OF ZERO CUSTOMERS IN THE SYSTEM = .5000
AVERAGE NUMBER OF CUSTOMERS IN THE SYSTEM = 19.2500
AVERAGE NUMBER OF CUSTOMERS IN THE QUEUE = 18.2500
AVERAGE WAITING-TIME IN THE SYSTEM = 3.85
AVERAGE WAITING-TIME IN THE QUEUE = 3.75

FIGURE 6-2: Sample outputs from Figure 6-1

ENTER AVERAGE CUSTOMER ARRIVAL RATE (LAMBDA)? 5
ENTER AVERAGE SERVICE RATE (MU)? 10
ENTER VARIANCE OF DISTRIBUTION? 1
PROBABILITY OF ZERO CUSTOMERS IN THE SYSTEM = .5000
AVERAGE NUMBER OF CUSTOMERS IN THE SYSTEM = 6.75
AVERAGE NUMBER OF CUSTOMERS IN THE QUEUE = 6.25
AVERAGE WAITING-TIME IN THE SYSTEM = 1.35
AVERAGE WAITING-TIME IN THE QUEUE = 1.25
ENTER AVERAGE CUSTOMER ARRIVAL RATE (LAMBDA)? 5
ENTER AVERAGE SERVICE RATE (MU)? 10
ENTER VARIANCE OF DISTRIBUTION? .5
PROBABILITY OF ZERO CUSTOMERS IN THE SYSTEM = .5000
AVERAGE NUMBER OF CUSTOMERS IN THE SYSTEM = 3.625
AVERAGE NUMBER OF CUSTOMERS IN THE QUEUE = 3.125
AVERAGE WAITING-TIME IN THE SYSTEM = .725
AVERAGE WAITING-TIME IN THE QUEUE = .625

ENTER AVERAGE CUSTOMER ARRIVAL RATE (LAMBDA)? 5
ENTER AVERAGE SERVICE RATE (MU)? 10
ENTER VARIANCE OF DISTRIBUTION? 0
PROBABILITY OF ZERO CUSTOMERS IN THE SYSTEM = .5000
AVERAGE NUMBER OF CUSTOMERS IN THE SYSTEM = .5
AVERAGE NUMBER OF CUSTOMERS IN THE QUEUE = 0
AVERAGE WAITING-TIME IN THE SYSTEM = .1
AVERAGE WAITING-TIME IN THE QUEUE = 0

FIGURE 6-2: (Continued)

The M/D/1 Queueing System

We shall now examine a single-server queueing system which follows a constant service-time probability distribution while the input process follows the Poisson probability distribution. The following assumptions are required in order that we may present this particular queueing system:

1. All service times for all possible customers are identically distributed.
2. The average customer arrival rate is denoted by λ.
3. There is a single server in the service facility.
4. The source population is infinite in size.
5. The queue discipline is FIFO.
6. The maximum allowable length of the queue is infinite.

Since the first assumption requires identical service times for all possible customers, this queueing system is best suited for those real-world situations where the same task is to be performed by the server on all customers, for example, an assembly line. In most assembly line situations, however, customer arrival rate is not a pure Poisson distribution. A much better real-world situation for the M/D/1 queueing system would be the beginning of an assembly line or a quality control area where the arriving customers follow a Poisson probability distribution while the single server performs the same task (tightening a bolt or stamping a package) on each and every arriving customer.

Since parameter λ defines the average customer arrival rate and parameter μ defines the average service rate, we may consider these average values in terms of the following:

$$\rho = \lambda/\mu < 1$$

It must be noted, of course, that the average service rate for all possible customers will equal the service rate for any single customer (constant service rate). Therefore, parameter λ becomes the deciding factor in the previous equation. This is not to say, however, that parameter μ will never change during the course of a particular study of the system. In fact, one of the primary reasons for examining the M/D/1 queueing system is to determine an acceptable level of service for the constant service time. If a rather long waiting line begins to form, given a particular value of μ, then the most obvious parameter to change would be the service rate value. This assumes, of course, that the operator of the queueing system has little or no control over the average customer arrival rate.

For the single-server service facility, the Poisson arrival and constant service-time queueing system becomes a special case of the Poisson arrival and any service-time queueing system. The equations which were used in the M/G/1 queueing system are simplified, since the variance of the constant service-time probability distribution is equal to zero. Therefore, we simply set $\sigma^2 = 0$ in the any-service-time equations in order to obtain the following:

$$L_q = \frac{\rho^2}{2(1 - \rho)} \tag{6-6}$$

$$L = \rho + L_q \tag{6-7}$$

$$W_q = L_q/\lambda \tag{6-8}$$

$$W = W_q + (1/\mu) \tag{6-9}$$

where $\rho = \lambda/\mu$, λ is the average customer arrival rate (a Poisson process), and μ is the average service rate (a constant).

The multiserver queueing system (M/D/c) is, unfortunately, a rather complicated system to model in an effective mathematical fashion. Prabhu[3] offers a thorough examination of the M/D/c queueing system.

---

[3]N.U. Prabhu, *Queues and Inventories* (New York: Wiley, 1965), pp. 32-34.

Figures 6-3 and 6-4 are the programmed routine and sample outputs, respectively, for the M/D/1 queueing system.

```
1000 REM    M/D/1 QUEUEING SYSTEM
1010 REM     VARIABLE-NAME      DESCRIPTION
1020 REM    L    AVERAGE NUMBER OF CUSTOMERS IN SYSTEM
1030 REM    LM   LAMBDA - AVERAGE CUSTOMER ARRIVAL RATE
1040 REM    LQ   AVERAGE NUMBER OF CUSTOMERS IN QUEUE
1050 REM    MU   MU - AVERAGE SERVICE RATE
1060 REM    RO   RHO - TRAFFIC INTENSITY RATIO
1070 REM    W    AVERAGE WAITING-TIME IN SYSTEM
1080 REM    WQ   AVERAGE WAITING-TIME IN QUEUE
1090 PRINT"ENTER AVERAGE CUSTOMER ARRIVAL RATE (LAMBDA)";
1100 INPUT LM
1110 PRINT"ENTER AVERAGE SERVICE RATE (MU)";
1120 INPUT MU
1130 IF LM/MU < 1.0 THEN 1180
1140   PRINT"TRAFFIC INTENSITY RATIO (LAMBDA/MU) MUST BE"
1150   PRINT"LESS THAN ONE."
1160   PRINT
1170   GOTO 1090
1180 RO = LM/MU
1190 LQ = (RO↑2)/(2*(1.0-RO))
1200 L = RO + LQ
1210 WQ = LQ/LM
1220 W = WQ + (1/MU)
1230 PRINT"AVERAGE NUMBER OF CUSTOMERS IN THE SYSTEM ="; L
1240 PRINT"AVERAGE NUMBER OF CUSTOMERS IN THE QUEUE ="; LQ
1250 PRINT"AVERAGE WAITING-TIME IN THE SYSTEM ="; W
1260 PRINT"AVERAGE WAITING-TIME IN THE QUEUE ="; WQ
1270 GOTO 1090
1280 END
```

FIGURE 6-3: Programmed routine for the M/D/1 queueing system

```
ENTER AVERAGE CUSTOMER ARRIVAL RATE (LAMBDA)? 5
ENTER AVERAGE SERVICE RATE (MU)? 10
AVERAGE NUMBER OF CUSTOMERS IN THE SYSTEM = .75
AVERAGE NUMBER OF CUSTOMERS IN THE QUEUE = .25
AVERAGE WAITING-TIME IN THE SYSTEM = .15
AVERAGE WAITING-TIME IN THE QUEUE = .05
```

FIGURE 6-4: Sample output from Figure 6-3

ENTER AVERAGE CUSTOMER ARRIVAL RATE (LAMBDA)? 5
ENTER AVERAGE SERVICE RATE (MU)? 6
AVERAGE NUMBER OF CUSTOMERS IN THE SYSTEM = 2.92
AVERAGE NUMBER OF CUSTOMERS IN THE QUEUE = 2.08
AVERAGE WAITING-TIME IN THE SYSTEM = .58
AVERAGE WAITING-TIME IN THE QUEUE = .42
ENTER AVERAGE CUSTOMER ARRIVAL RATE (LAMBDA)? 10
ENTER AVERAGE SERVICE RATE (MU)? 20
AVERAGE NUMBER OF CUSTOMERS IN THE SYSTEM = .75
AVERAGE NUMBER OF CUSTOMERS IN THE QUEUE = .25
AVERAGE WAITING-TIME IN THE SYSTEM = .08
AVERAGE WAITING-TIME IN THE QUEUE = .03

FIGURE 6-4: (Continued)

Although an assembly-line situation is often considered to be a constant-service rate queueing system, we shall examine the particular case which may be realistically modeled by the operation of a fast-food business employing a single cook. The single server here is the cook. Customer arrivals will be considered to have a Poisson probability distribution with an average arrival rate equal to parameter λ. The service time probability distribution will assume a constant value, that is, the cook does nothing but grill the infamous six-minute hamburger.

For our purposes, we shall initially assume that the constant service-time rate (μ) is equal to six while the average customer arrival rate (λ) is equal to four. Under these conditions, the following solutions are given for the M/D/1 queueing system modeled in the programmed routing of Figure 6-3:

$$
\begin{aligned}
L &= 1.3333 \\
L_q &= 0.6667 \\
W &= 0.3333 \\
W_q &= 0.1687
\end{aligned}
$$

If the customer arrival rate is given in customers per hour, the customer will be required to wait in the system for 20 minutes in order to receive a six-minute hamburger. This waiting time is, of course, subject to the previous input parameters as given.

There is very little we can do with this queueing system, since the service-time parameter is considered to be not only constant but fixed. We may attempt to speed up the cooking process, however this adjustment may not be allowable when the requirements of the service are of the constant variety. The customer arrival rate may be adjusted, but this parameter is not under the control of the person modeling this queueing system. In fact, customer arrivals are completely dependent upon the whims of the source population and we may study only the overall effects of various customer arrivals.

The reader is urged to input various other values for the customer arrival parameter in order to examine the effects, if any, which the lunch hour crowd may have on the operation of this establishment.

The examination of constant service-time queueing systems is rather limited considering the real-world queueing systems which fail to fall into this particular category. However, study of this system is included as a brief introduction to the categories of queueing systems which inhabit our lives.

The D/D/1 Queueing System

When both the customer arrival rate and the service rate of a single-server queueing system are constant, we have a D/D/1 queueing system. In this case, the average customer arrival rate (λ) and the average service rate (μ) are considered essentially to be constant over the period of time when this queueing system is being studied. Our previous assumptions regarding the various components of the system also hold true—the source population is considered to be infinite in size, the allowable queue length is assumed to be infinite in size, and the queue discipline is of an FIFO nature.

Given the input parameters λ and μ, we may also assume that $\rho = \lambda/\mu < 1$. The following queue measurement equations hold for the D/D/1 queueing system:

$$P_0 = 1 - (\lambda/\mu)$$
$$P_n = (\lambda/\mu)^n P_0$$

The probability of n customers being in the system may also be stated as

$$P_n = (1 - \rho)\rho^n \tag{6-10}$$

where $n = 0, 1, 2, \ldots$ In addition, the following queue measurements will be used to effectively examine the D/D/1 queueing system:

$$L = \lambda/(\mu - \lambda) \tag{6-11}$$
$$L_q = \lambda^2/(\mu(\mu - \lambda)) \tag{6-12}$$

When the average customer arrival rate (λ) is greater than the average service rate (μ), the above equations as well as the waiting line grow without bound. Assuming $\lambda < \mu$,

$$W = \mu/(\mu - \lambda) \tag{6-13}$$
$$W_q = \lambda/(\mu(\mu - \lambda)) \tag{6-14}$$

Figure 6-5 is the programmed routine which supplies the sample output for Figure 6-6.

```
1000 REM    D/D/1 QUEUEING SYSTEM
1010 REM    VARIABLE-NAME    DESCRIPTION
1020 REM    L    AVERAGE NUMBER OF CUSTOMERS IN SYSTEM
1030 REM    LM   LAMBDA - AVERAGE CUSTOMER ARRIVAL RATE
1040 REM    LQ   AVERAGE NUMBER OF CUSTOMERS IN QUEUE
1050 REM    MU   MU - AVERAGE SERVICE RATE
1060 REM    N    CUSTOMER COUNTER
1070 REM    P(0) PROBABILITY OF 0 CUSTOMERS IN SYSTEM
1080 REM    P(N) PROBABILITY OF N CUSTOMERS IN SYSTEM
1090 REM    RO   RHO - TRAFFIC INTENSITY RATIO
1100 REM    W    AVERAGE WAITING-TIME IN SYSTEM
1110 REM    WQ   AVERAGE WAITING-TIME IN QUEUE
1120 DIM P(5)
1130 PRINT"ENTER AVERAGE CUSTOMER ARRIVAL RATE (LAMBDA)";
1140 INPUT LM
1150 PRINT"ENTER AVERAGE SERVICE RATE (MU)";
1160 INPUT MU
1170 IF LM/MU < 1.0 THEN 1240
1180   PRINT"TRAFFIC INTENSITY RATIO (LAMBDA/MU) MUST BE"
1190   PRINT"LESS THAN ONE IF THE PROBABILITY OF N CUSTOMERS"
1200   PRINT"IN THE SYSTEM AT TIME T IS TO BE INDEPENDENT"
1210   PRINT"OF THE TIME T."
1220   PRINT
1230   GOTO 1090
1240 RO = LM/MU
1250 FOR N = 0 TO 5
1260 P(N) = (1.0 - RO) * (RO↑N)
1270 NEXT N
1280 L = LM/(MU-LM)
1290 LQ = (LM ↑ 2)/(MU * (MU - LM))
1300 W = MU/(MU - LM)
1310 WQ = LM/(MU * (MU - LM))
1320 PRINT"PROBABILITY (P) OF (N) CUSTOMERS IN THE SYSTEM"
1330 PRINT"(N)          (P)"
1340 FOR N = 0 TO 5
1350   PRINT N, P(N)
```

FIGURE 6-5: Programmed routine for the D/D/1 queueing system

1360 NEXT N
1370 PRINT"AVERAGE NUMBER OF CUSTOMERS IN THE SYSTEM ="; L
1380 PRINT"AVERAGE NUMBER OF CUSTOMERS IN THE QUEUE ="; LQ
1390 PRINT"AVERAGE WAITING-TIME IN THE SYSTEM = "; W
1400 PRINT"AVERAGE WAITING-TIME IN THE QUEUE ="; WQ
1410 GOTO 1130
1420 END

FIGURE 6-5: (Continued)

ENTER AVERAGE CUSTOMER ARRIVAL RATE (LAMBDA)? 5
ENTER AVERAGE SERVICE RATE (MU)? 10
PROBABILITY (P) OF (N) CUSTOMERS IN THE SYSTEM

(N)	(P)
0	.5000
1	.2500
2	.1250
3	.0625
4	.0313
5	.0156

AVERAGE NUMBER OF CUSTOMERS IN THE SYSTEM = 1.0
AVERAGE NUMBER OF CUSTOMERS IN THE QUEUE = 0.5
AVERAGE WAITING-TIME IN THE SYSTEM = 2.0
AVERAGE WAITING-TIME IN THE QUEUE = 0.1

ENTER AVERAGE CUSTOMER ARRIVAL RATE (LAMBDA)? 10
ENTER AVERAGE SERVICE RATE (MU)? 20
PROBABILITY (P) OF (N) CUSTOMERS IN THE SYSTEM

(N)	(P)
0	.5000
1	.2500
2	.1250
3	.0625
4	.0313
5	.0156

AVERAGE NUMBER OF CUSTOMERS IN THE SYSTEM = 1.0
AVERAGE NUMBER OF CUSTOMERS IN THE QUEUE = 0.5
AVERAGE WAITING-TIME IN THE SYSTEM = 2.0
AVERAGE WAITING-TIME IN THE QUEUE = 0.05

FIGURE 6-6: Sample output from figure 6-5

ENTER AVERAGE CUSTOMER ARRIVAL RATE (LAMBDA)? 1
ENTER AVERAGE SERVICE RATE (MU)? 5
PROBABILITY (P) OF (N) CUSTOMERS IN THE SYSTEM

(N)	(P)
0	.8000
1	.1600
2	.0320
3	.0064
4	.0013
5	.0003

AVERAGE NUMBER OF CUSTOMERS IN THE SYSTEM = .25
AVERAGE NUMBER OF CUSTOMERS IN THE QUEUE = .05
AVERAGE WAITING-TIME IN THE SYSTEM = 1.25
AVERAGE WAITING-TIME IN THE QUEUE = 0.05

FIGURE 6-6: (Continued)

Problems for Reader Solution

6-1. Use equation 6-1 to calculate the value for P_0 given the following input parameters (variance = 0.05):
   a) $\mu = 10$ and $\lambda = 8$
   b) $\mu = 10$ and $\lambda = 9$

6-2. Use equations 6-2 through 6-5 to calculate the values for L, L_q, W, and W_q given the following input parameters (variance = 0.05):
   a) $\mu = 10$ and $\lambda = 8$
   b) $\mu = 10$ and $\lambda = 9$

6-3. Use equations 6-6 through 6-9 to calculate the values for L, L_q, W, and W_q given the following input parameters:
   a) $\mu = 10$ and $\lambda = 8$
   b) $\mu = 10$ and $\lambda = 9$

6-4. Use equation 6-10 to calculate the values for P_n (where $n = 0$, 1, and 2) given the following input parameters:
   a) $\mu = 10$ and $\lambda = 8$
   b) $\mu = 10$ and $\lambda = 9$

6-5. Use equations 6-11 through 6-14 to calculate the values for L, L_q, W, and W_q given the following input parameters:
   a) $\mu = 10$ and $\lambda = 8$
   b) $\mu = 10$ and $\lambda = 9$

Answers to Problems

6-1.
   a) 0.200
   b) 0.100

6-2.
   a) $L = 10.400, L_q = 9.600, W = 1.300, W_q = 1.200$
   b) $L = 25.200, L_q = 24.300, W = 2.700, W_q = 2.800$

6-3.
   a) $L = 2.40, L_q = 1.60, W = 0.30, W_q = 0.20$
   b) $L = 4.95, L_q = 4.05, W = 0.55, W_q = 0.45$

6-4.
   a) $P_0 = 0.200, P_1 = 0.160, P_2 = 0.128$
   b) $P_0 = 0.100, P_1 = 0.090, P_2 = 0.081$

6-5.
   a) $L = 4.0, L_q = 3.2, W = 5.0, W_q = 0.4$
   b) $L = 9.0, L_q = 8.1, W = 10.0, W_q = 0.9$

7

The End of the Line

Finally, we have attained our goal—the end of our waiting line. It has been a long and, at times, an arduous journey. The information presented in this book will provide those of us who wait in lines a more meaningful description of queueing systems. Perhaps, our wait will be more interesting next time as we look at waiting lines rather than just be a spectator in the waiting line.

Before we leave this last service facility, however, we should become acquainted with the summary information provided by the appendices. In particular, Appendix A lists the formulas which we examined during our investigation of probability theory. Appendix B, which lists values for the exponential factor, may prove useful if we do not have a computer available. Appendix C lists values for a typical exponential process given various values for lambda. The notational glossary in Appendix D summarizes some of the more useful queueing system terms. Appendices E, F, and G list the appropriate equations for the typical queueing system (M/M/1, M/M/C, M/E$_k$/1, respectively). Appendices H, I, and J do the same for the M/G/1, the M/D/1, and the D/D/1 queueing systems, respectively. Appendix K is a summary programmed routine of the various queueing systems examined in previous chapters. This overall routine may prove more useful than the fragmented routines presented in each chapter. The sample output provided by this programmed routine should be used to check the "keypunching" of the routine into your particular system.

The Bibliography contains a wealth of references useful to the reader interested in a further study of queueing systems.

145

A. Probability Theory Formulas

DISTRIBUTION	PROBABILITY DENSITY FUNCTION	MEAN VALUE	VARIANCE
Binomial	$\binom{n}{x} p^x q^{n-x}$	np	npq
Erlang-k	$(k\lambda)^k \dfrac{\exp^{-k\lambda x}}{(k-1)!} x^{k-1}$		μ^2/k
Exponential	$\lambda \exp^{-\lambda x}$	$1/\lambda$	$1/\lambda^2$
Gamma	$\dfrac{a^k x^{(k-1)} \exp^{-ax}}{(k-1)!}$	k/a	k/a^2
Normal	$\dfrac{1}{(2\pi\sigma)^{1/2}} \exp^{-(x-\mu)^2/2\sigma^2}$	μ	σ^2
Poisson	$\exp^{-\lambda} \dfrac{\lambda^x}{x!}$	λ	λ

B. Values for the Exponential Factor

$$exp = 2.71828183 \ldots$$
$$lg \, exp = 0.43429448 \ldots$$
$$1/exp = 0.36787944 \ldots$$
$$lg \, (1/exp) = 0.56570552 \ldots -1.0$$

x	exp^x	exp^{-x}
0.00	1.0000	1.0000
0.01	1.0101	0.9900
0.02	1.0202	0.9802
0.03	1.0305	0.9704
0.04	1.0408	0.9608
0.05	1.0513	0.9512
0.06	1.0618	0.9418
0.07	1.0725	0.9324
0.08	1.0833	0.9231
0.09	1.0942	0.9139
0.10	1.1052	0.9048
0.11	1.1163	0.8958
0.12	1.1275	0.8869
0.13	1.1388	0.8781
0.14	1.1503	0.8694
0.15	1.1618	0.8607
0.16	1.1735	0.8521
0.17	1.1853	0.8437
0.18	1.1972	0.8353

x	exp^x	exp^{-x}
0.19	1.2092	0.8270
0.20	1.2214	0.8187
0.21	1.2337	0.8106
0.22	1.2461	0.8025
0.23	1.2586	0.7945
0.24	1.2712	0.7866
0.25	1.2840	0.7788
0.26	1.2969	0.7711
0.27	1.3100	0.7634
0.28	1.3231	0.7558
0.29	1.3364	0.7483
0.30	1.3499	0.7408
0.31	1.3634	0.7334
0.32	1.3771	0.7261
0.33	1.3910	0.7189
0.34	1.4049	0.7118
0.35	1.4191	0.7047
0.36	1.4333	0.6977
0.37	1.4477	0.6907
0.38	1.4623	0.6839
0.39	1.4770	0.6771
0.40	1.4918	0.6703
0.41	1.5068	0.6637
0.42	1.5220	0.6570
0.43	1.5373	0.6505
0.44	1.5527	0.6440
0.45	1.5683	0.6376
0.46	1.5841	0.6313
0.47	1.6000	0.6250
0.48	1.6161	0.6188
0.49	1.6323	0.6126
0.50	1.6487	0.6065
0.60	1.8221	0.5488
0.70	2.0138	0.4966
0.80	2.2255	0.4493
0.90	2.4596	0.4066
1.00	2.7183	0.3679
2.00	7.3891	0.1353

x	exp^x	exp^{-x}
3.00	20.086	0.0498
4.00	54.598	0.0183
5.00	148.41	0.0067
6.00	403.43	0.0025
7.00	1096.6	0.0009
8.00	2981.0	0.0003
9.00	8103.1	0.0001

C. Values for the Exponential Process

λ x	0	1	2	3	4	5	6	7	8	9
0	0	0	0	0	0	0	0	0	0	0
1	0	.6321	.8647	.9502	.9817	.9933	.9975	.9991	.9997	.9999
2	0	.8647	.9817	.9975	.9997	.9999	.9999	.9999	.9999	.9999
3	0	.9502	.9975	.9999	.9999	.9999	.9999	.9999	.9999	.9999
4	0	.9817	.9997	.9999	.9999	.9999	.9999	.9999	.9999	.9999
5	0	.9933	.9999	.9999	.9999	.9999	.9999	.9999	.9999	.9999
6	0	.9975	.9999	.9999	.9999	.9999	.9999	.9999	.9999	.9999
7	0	.9991	.9999	.9999	.9999	.9999	.9999	.9999	.9999	.9999
8	0	.9997	.9999	.9999	.9999	.9999	.9999	.9999	.9999	.9999
9	0	.9999	.9999	.9999	.9999	.9999	.9999	.9999	.9999	.9999

D. *Glossary of Notations*

A/B/c	Kendall queueing system notation
A/B/c/K/m/Z	Extended Kendall queueing notation
A	Customer arrival probability distribution
B	Service-time probability distribution
c	Number of identical servers in service facility
C_x^2	Squared coefficient of variation
D	Deterministic (constant) interarrival-time or service-time probability distribution
E_k	Erlang-k interarrival-time or service-time probability distribution
FCFS	First-come first-served queue discipline
FIFO	First-in-first-out queue discipline (identical to FCFS queue discipline)
$f(x)$	Probability density function
$F(x)$	Cumulative distribution function
G	General service-time probability distribution
GI	General independent interarrival-time probability distribution
K	Maximum number of allowable customers into the queueing system
k	Order of an Erlang-k probability distribution
L	Average number of customers in the queueing system
L_q	Average number of customers in the waiting line
LCFS	Last-come first-served queue discipline
LIFO	Last-in first-out queue discipline (identical to LCFS queue discipline

λ	Average customer arrival rate
λ'	Traffic rate of arriving customers
M	Exponential interarrival-time or service-time probability distribution
m	Number of customers in a finite source population
n	Number of customers in a queueing system
σ	Standard deviation
σ^2	Variance
μ	Average service rate per server
μ	Mean or expected value
$P_0(t)$	Probability of zero customers in the system at time t
$P_n(t)$	Probability of n customers in the system at time t
ρ	Server utilization ratio
SIRO	Service-in-random-order queue discipline
t	time period
υ	Traffic intensity ratio in erlangs
W	Average waiting time in the system
W_q	Average waiting time in the queue
X	Random variable
x	Random variate value
Z	Type of queue discipline

E. M/M/1 Queueing System Equations

1. Probability of zero customers in the system at time t:

$$P_0(t) = 1 - (\lambda/\mu) = 1 - \rho$$

2. Probability of n customers in the system at time t:

$$P_n(t) = \rho^n P_0(t) = \rho^n (1-\rho)$$

3. Average number of customers in the system:

$$L = \lambda/(\mu - \lambda)$$

4. Average number of customers in the waiting line:

$$L_q = \lambda^2 /(\mu(\mu - \lambda))$$

5. Average waiting time in the system:

$$W = 1/(\mu - \lambda)$$

6. Average waiting time in the queue:

$$W_q = \lambda/(\mu(\mu - \lambda))$$

F. M/M/c Queueing System Equations

1. Probability of zero customers in the system at time t:

$$P_0(t) = \cfrac{1}{\cfrac{\displaystyle\sum_{n=0}^{c-1}\left(\dfrac{\lambda}{\mu}\right)^n}{n!} + \cfrac{\left(\dfrac{\lambda}{\mu}\right)^c}{c!(1-(\rho/c))}}$$

2. Probability of n customers in the system when $n < c$:

$$P_n(t) = (1/n!)\rho^n P_0(t)$$

3. Probability of n customers in the system when $n \geq c$:

$$P_n(t) = 1/(c!c^{n-c})\rho^n P_0(t)$$

4. Average waiting time in the queue:

$$W_q = \cfrac{\cfrac{\rho^c P_0(t)}{c!(1-(\rho/c))}}{c\mu(1-(\rho/c))}$$

5. Average waiting time in the system:

$$W = W_q + 1/\mu$$

6. Average number of customers in the waiting line:

$$L_q = \lambda W_q$$

7. Average number of customers in the system:

$$L = \lambda W$$

G. $M/E_k/1$ Queueing System Equations

1. Average number of customers in the waiting line:

$$L_q = \frac{k+1}{2k} \times \frac{\lambda^2}{\mu(\mu - \lambda)}$$

2. Average number of customers in the system:

$$L = \frac{k+1}{2k} \times \frac{\lambda^2}{\mu(\mu - \lambda)} + \rho$$

3. Average waiting time in the queue:

$$W_q = \frac{k+1}{2k} \times \frac{\lambda}{\mu(\mu - \lambda)}$$

4. Average waiting time in the system:

$$W = \frac{k+1}{2k} \times \frac{\lambda}{\mu(\mu - \lambda)} + (1/\mu)$$

H. $M/G/1$ *Queueing System Equations*

1. Probability of zero customers in the system at time t:

$$P_0(t) = 1 - (\lambda/\mu)$$

2. Average number of customers in the waiting line:

$$L_q = \frac{\lambda^2 \sigma^2 + \rho^2}{2(1 - \rho)}$$

3. Average number of customers in the system:

$$L = \rho + L_q$$

4. Average waiting time in the queue:

$$W_q = L_q/\lambda$$

5. Average waiting time in the system:

$$W = L/\lambda$$

I. M/D/1 Queueing System Equations

1. Average number of customers in the waiting line:

$$L_q = \frac{\rho^2}{2(1 - \rho)}$$

2. Average number of customers in the system:

$$L = \rho + L_q$$

3. Average waiting time in the queue:

$$W_q = L_q/\lambda$$

4. Average waiting time in the system:

$$W = W_q + (1/\mu)$$

J. D/D/1 Queueing System Equations

1. Probability of n customers in the system at time t for

$$n = 0, 1, 2, \ldots :$$
$$P_n = (1 - \rho)\rho^n$$

2. Average number of customers in the system:

$$L = \lambda/(\mu - \lambda)$$

3. Average number of customers in the waiting line:

$$L_q = \frac{\lambda^2}{\mu(\mu - \lambda)}$$

4. Average waiting time in the queue:

$$W_q = \frac{\lambda}{\mu(\mu - \lambda)}$$

5. Average waiting time in the system:

$$W = \mu/(\mu - \lambda)$$

K. General Programmed Routine for Queueing Systems

```
1000  REM* * * * * QUEUEING SYSTEMS * * * * *
1010  REM              LEONARD GORNEY
1020  REM      VARIABLE-NAME      DESCRIPTION
1030  REM      A$     RERUN SAME QUEUE-TYPE IF = "Y"
1040  REM      C      NUMBER OF SERVERS
1050  REM      K      ERLANG-K PARAMETER
1060  REM      L      AVERAGE # OF CUSTOMERS IN SYSTEM
1070  REM      LM     AVERAGE CUSTOMER ARRIVAL RATE
1080  REM      LQ     AVERAGE # OF CUSTOMERS IN QUEUE
1090  REM      MU     AVERAGE SERVICE RATE
1100  REM      N      CUSTOMER COUNTER
1110  REM      NC     USED FOR FACTORIAL CALCULATION
1120  REM      NF     USED FOR FACTORIAL CALCULATION
1130  REM      NM     USED FOR FACTORIAL CALCULATION
1140  REM      NS     USED FOR FACTORIAL CALCULATION
1150  REM      P      PROBABILITY OF N CUSTOMERS
1160  REM      PN     TEMPORARY PROBABILITY OF N CUSTOMERS
1170  REM      PS     TEMPORARY PROBABILITY OF N CUSTOMERS
1180  REM      RO     TRAFFIC INTENSITY RATIO
1190  REM      V2     VARIANCE OF DISTRIBUTION
1200  REM      W      AVERAGE WAIT-TIME IN SYSTEM
1210  REM      WQ     AVERAGE WAIT-TIME IN QUEUE
1220  DIM P(5)
1230  PRINT"QUEUE-TYPE"
1240  PRINT"    1     M/M/1"
1250  PRINT"    2     M/M/C"
```

```
1260 PRINT"     3    M/E(K)/1"
1270 PRINT"     4    M/G/1"
1280 PRINT"     5    M/D/1"
1290 PRINT"     6    D/D/1"
1300 INPUT"QUEUE-TYPE"; QT
1310 ON QT GOSUB 1350, 1490, 1850, 1950, 2070, 2160
1320 INPUT"ENTER 'Y' FOR SAME QUEUE-TYPE ELSE ENTER 'N'"; A$
1330 IF A$ = "Y" THEN 1310
1340    GOTO 1230
1350 REM * * * * * M / M / 1 * * * * *
1360 GOSUB 2290
1370 P(0) = 1.0 - RO
1380 FOR N = 1 TO 5
1390   P(N) = (RO ↑ N) * (1.0 - RO)
1400 NEXT N
1410 L = LM/(MU - LM)
1420 LQ = (LM ↑ 2)/(MU * (MU - LM))
1430 W = 1.0/(MU - LM)
1440 WQ = LM/(MU * (MU - LM))
1450 GOSUB 2370
1460 GOSUB 2440
1470 GOSUB 2560
1480 RETURN
1490 REM * * * * * M / M / C * * * * *
1500 GOSUB 2290
1510 INPUT"ENTER NUMBER OF SERVERS (C)"; C
1520 RO = LM/(MU * C)
1530 P(0) = 0.0
1540 NM = C
1550 GOSUB 2500
1560 PS = (1.0/NF) * ((LM/MU)↑C) * ((C*MU)/((C*MU)-LM))
1570 PN = 0.0
1580 FOR NS = 1 TO (C-1)
1590   NM = NS
1600   GOSUB 2500
1610   PN = (1.0/NF) * ((LM/MU)↑NS) + PN
1620 NEXT NS
1630 P(0) = PN + PS
1640 FOR NS = 1 TO 5
1650   IF NS < C THEN 1700
1660     NM = C
1670     GOSUB 2500
```

```
1680    P(NS) = 1.0 / (NF*C↑(NS-C)) * RO↑NS * P(0)
1690    GOTO 1730
1700    NM = NS
1710    GOSUB 2500
1720    P(NS) = (1.0/NF) * RO↑NS * P(0)
1730 NEXT NS
1740 NM = C - 1
1750 GOSUB 2500
1760 L = ((MU*LM*(RO↑C)) / (NF*(((C*MU)-LM)↑2)) * P(0)
1770 LQ = L + RO
1780 W = ((MU*(RO↑C)) / ( NF * ((C*MU)-LM)↑2)) * P(0)
1790 WQ = WQ + (1.0/MU)
1800 GOSUB 2370
1810 GOSUB 2440
1820 GOSUB 2560
1830 PRINT"SERVER UTILIZATION RATIO ="; LM / (MU * C)
1840 RETURN
1850 REM * * * * * M / E(K) / 1 * * * * *
1860 GOSUB 2290
1870 INPUT"ENTER ERLANG PARAMETER (K)"; K
1880 LQ = (((K+1)/2)*K)*((LM↑2)/(MU*(MU-LM)))
1890 L = LQ + RO
1900 WQ = (((K+1)/2)*K)*(LM/(MU*(MU-LM)))
1910 W = WQ + ( 1.0 / MU )
1920 GOSUB 2440
1930 GOSUB 2560
1940 RETURN
1950 REM * * * * * M / G / 1 * * * * *
1960 GOSUB 2290
1970 INPUT"ENTER VARIANCE OF SERVICE DISTRIBUTION"; V2
1980 P(0) = 1.0 - RO
1990 LQ - (((LM↑2)*V2)+(RO↑2))/(2*(1.0-RO))
2000 L = RO + LQ
2010 W = LQ / LM
2020 WQ = L / LM
2030 PRINT"PROBABILITY OF ZERO CUSTOMERS IN SYSTEM ="; P(0)
2040 GOSUB 2440
2050 GOSUB 2560
2060 RETURN
2070 REM * * * * * M / D / 1 * * * * *
2080 GOSUB 2290
2090 LQ = ( RO ↑ 2 ) / ( 2 * ( 1.0 - RO ))
```

```
2100  L = RO + LQ
2110  WQ = LQ / LM
2120  W = WQ + (1.0 / MU )
2130  GOSUB 2440
2140  GOSUB 2560
2150  RETURN
2160  REM * * * * * D / D / 1 * * * * *
2170  GOSUB 2290
2180  FOR N = 0 TO 5
2190  P(N) = ( 1.0 - RO ) * ( RO ↑ N )
2200  NEXT N
2210  L = LM / ( MU - LM )
2220  LQ = (LM ↑ 2)/(MU * (MU - LM))
2230  W = MU / ( MU - LM )
2240  WQ = LM / ( MU * (MU - LM ))
2250  GOSUB 2370
2260  GOSUB 2440
2270  GOSUB 2560
2280  RETURN
2290  REM * * * * * INPUT PARAMETERS * * * * *
2300  INPUT"ENTER AVERAGE SERVICE RATE (MU)"; MU
2310  INPUT"ENTER AVERAGE CUSTOMER
          ARRIVAL RATE (LAMBDA)"; LM
2320  IF LM < MU THEN 2350
2330    PRINT"LAMBDA MUST BE LESS THAN MU"
2340    GOTO 2300
2350  RO = LM / MU
2360  RETURN
2370  REM * * * * * OUTPUT P(N) WHERE N = 0, 1, 2, . . . * * * * *
2380  PRINT"PROBABILITY (P) OF (N) CUSTOMERS"
2390  PRINT"(N)              (P)"
2400  FOR N = 0 TO 5
2410    PRINT N, P(N)
2420  NEXT N
2430  RETURN
2440  REM * * * * * OUTPUT L, LQ, W, WQ * * * * *
2450  PRINT"AVERAGE NUMBER OF CUSTOMERS IN SYSTEM ="; L
2460  PRINT"AVERAGE NUMBER OF CUSTOMERS IN QUEUE ="; LQ
2470  PRINT"AVERAGE WAIT-TIME TO COMPLETE SERVICE ="; W
2480  PRINT"AVERAGE WAIT-TIME IN WAITING-LINE ="; WQ
2490  RETURN
2500  REM * * * * * CALCULATE N FACTORIAL * * * * *
2510  NF = 1
```

```
2520  FOR NC = NM TO 1 STEP –1
2530    NF = NF * NC
2540  NEXT NC
2550  RETURN
2560  REM * * * * * OUTPUT TRAFFIC INTENSITY RATIO * * * * *
2570  PRINT"TRAFFIC INTENSITY RATIO ="; LM / MU
2580  RETURN
2590  END
```

Sample Outputs

QUEUE–TYPE
    1 M/M/1
    2 M/M/C
    3 M/E(K)/1
    4 M/G/1
    5 M/D/1
    6 D/D/1
QUEUE–TYPE? 1
ENTER AVERAGE SERVICE RATE (MU)? 10
ENTER AVERAGE CUSTOMER ARRIVAL RATE (LAMBDA)? 5
PROBABILITY (P) OF (N) CUSTOMERS

(N)	(P)
0	.5
1	.25
2	.125
3	.0625
4	.03125
5	.015625

AVERAGE NUMBER OF CUSTOMERS IN SYSTEM = 1
AVERAGE NUMBER OF CUSTOMERS IN QUEUE = .5
AVERAGE WAIT–TIME TO COMPLETE SERVICE = .2
AVERAGE WAIT–TIME IN WAITING–LINE = .1
TRAFFIC INTENSITY RATIO = .5
ENTER 'Y' FOR SAME QUEUE–TYPE ELSE ENTER 'N'? N
QUEUE–TYPE
    1 M/M/1
    2 M/M/C
    3 M/E(K)/1
    4 M/G/1
    5 M/D/1
    6 D/D/1

QUEUE-TYPE? <u>2</u>
ENTER AVERAGE SERVICE RATE (MU)? <u>12</u>
ENTER AVERAGE CUSTOMER ARRIVAL RATE (LAMBDA)? <u>22</u>
ENTER NUMBER OF SERVERS (C)? <u>2</u>
PROBABILITY (P) OF (N) CUSTOMERS

(N)	(P)
0	.0455
1	.0833
2	.0764
3	.0700
4	.0642
5	.0588

AVERAGE NUMBER OF CUSTOMERS IN SYSTEM = 11.9167
AVERAGE NUMBER OF CUSTOMERS IN QUEUE = 10.0833
AVERAGE WAITING-TIME TO COMPLETE SERVICE = .5417
AVERAGE WAITING-TIME IN WAITING-LINE = .4583
TRAFFIC INTENSITY RATIO = 1.8333
SERVER UTILIZATION RATIO = .9167
ENTER 'Y' FOR SAME QUEUE-TYPE ELSE ENTER 'N'? <u>N</u>
QUEUE-TYPE
   1 M/M/1
   2 M/M/C
   3 M/E(K)/1
   4 M/G/1
   5 M/D/1
   6 D/D/1
QUEUE-TYPE? <u>3</u>
ENTER AVERAGE SERVICE RATE (MU)? <u>10</u>
ENTER AVERAGE CUSTOMER ARRIVAL RATE (LAMBDA)? <u>5</u>
ENTER ERLANG PARAMETER (K)? <u>2</u>
AVERAGE NUMBER OF CUSTOMERS IN SYSTEM = 2
AVERAGE NUMBER OF CUSTOMERS IN QUEUE = 1.5
AVERAGE WAIT-TIME TO COMPLETE SERVICE = .4
AVERAGE WAIT-TIME IN WAITING-LINE = .3
TRAFFIC INTENSITY RATIO = .5
ENTER 'Y' FOR SAME QUEUE-TYPE ELSE ENTER 'N'? <u>N</u>
QUEUE-TYPE
   1 M/M/1
   2 M/M/C
   3 M/E(K)/1
   4 M/G/1
   5 M/D/1
   6 D/D/1

QUEUE–TYPE? 4
ENTER AVERAGE SERVICE RATE (MU)? 10
ENTER AVERAGE CUSTOMER ARRIVAL RATE (LAMBDA)? 5
ENTER VARIANCE OF SERVICE DISTRIBUTION? 2
PROBABILITY OF ZERO CUSTOMERS IN SYSTEM = .5
AVERAGE NUMBER OF CUSTOMERS IN SYSTEM = 50.75
AVERAGE NUMBER OF CUSTOMERS IN QUEUE = 50.25
AVERAGE WAIT–TIME TO COMPLETE SERVICE = 10.05
AVERAGE WAIT–TIME IN WAITING–LINE = 10.15
TRAFFIC INTENSITY RATIO = .5
ENTER 'Y' FOR SAME QUEUE–TYPE ELSE ENTER 'N'? N
QUEUE–TYPE

    1 M/M/1
    2 M/M/C
    3 M/E(K)/1
    4 M/G/1
    5 M/D/1
    6 D/D/1

QUEUE–TYPE? 5
ENTER AVERAGE SERVICE RATE (MU)? 10
ENTER AVERAGE CUSTOMER ARRIVAL RATE (LAMBDA)? 5
AVERAGE NUMBER OF CUSTOMERS IN SYSTEM = .75
AVERAGE NUMBER OF CUSTOMERS IN QUEUE = .25
AVERAGE WAIT–TIME TO COMPLETE SERVICE = .15
AVERAGE WAIT–TIME IN WAITING–LINE = .05
TRAFFIC INTENSITY RATIO = .5
ENTER 'Y' FOR SAME QUEUE–TYPE ELSE ENTER 'N'? N
QUEUE–TYPE

    1 M/M/1
    2 M/M/C
    3 M/E(K)/1
    4 M/G/1
    5 M/D/1
    6 D/D/1

QUEUE–TYPE? 6
ENTER AVERAGE SERVICE RATE (MU)? 10
ENTER AVERAGE CUSTOMER ARRIVAL RATE (LAMBDA)? 5
PROBABILITY (P) OF (N) CUSTOMERS

(N)	(P)
0	.5
1	.25
2	.125
3	.0625

4   .03125
5   .015625
AVERAGE NUMBER OF CUSTOMERS IN SYSTEM = 1
AVERAGE NUMBER OF CUSTOMERS IN QUEUE = .5
AVERAGE WAIT-TIME TO COMPLETE SERVICE = 2
AVERAGE WAIT-TIME IN WAITING-LINE = .1
TRAFFIC INTENSITY RATIO = .5

Bibliography

Allen, A.O. "Elements of Probability for System Design." *IBM Systems Journal* 13 (1974): 325-48.

——. "Elements of Queueing Theory for System Design." *IBM Systems Journal* 14 (1975): 161-87.

——. *Probability, Statistics, and Queueing Theory with Computer Applications.* New York: Academic Press, 1978.

——. "Queueing Models of Computer Systems." *Computer* 13 (1980): 13-24.

Beckmann, P. *Introduction to Elementary Queueing Theory and Telephone Traffic.* Boulder, CO: Golem Press, 1968.

Benes, V.E. *General Stochastic Processes in the Theory of Queues.* Reading, MA: Addison-Wesley, 1963.

——. "On Queues with Poisson Arrivals." *Annals of Mathematical Statistics* 28 (1956): 670-77.

Bharucha-Reid, A.T. *Elements of the Theory of Markov Processes and Their Applications.* New York: McGraw-Hill, 1960.

Bhat, U.N.; Shalaby, M.; and Fischer, M.J. "Approximation Techniques in the Solution of Queueing Problems." *Naval Research Logistics Quarterly* 26 (1979): 311-26.

Buzen, J.P., and Denning, P.J. "Measuring and Calculating Queue Length Distributions." *Computer* 13 (1980): 33-44.

Chang, W. "Preemptive Priority Queues." *Operations Research* 13 (1965): 620-23.

——. "Queueing with Nonpreemptive and Preemptive-resume Priorities." *Operations Research* 13 (1965): 1020-22.

——. "Single Server Queueing Processes in Computing Systems." *IBM Systems Journal* 9 (1970): 36-71.

Clarke, A.B., and Disney, R.L. *Probability and Random Processes for Engineers and Scientists.* New York: Wiley, 1970.

Cobham, A. "Priority Assignment in Waiting Line Problems." *Operations Research* 2 (1954): 70-76.

——. "Priority Assignment—A Correction." *Operations Research* 3 (1955): 547.

Cohen, J.W. *The Single Server Queue.* Amsterdam: North-Holland, 1969.

Cooper, R.B. *Introduction to Queueing Theory.* New York: Macmillan, 1972.

Cox, D.R., and Smith, W.L. *Queues.* New York: Wiley, 1963.

Cox, D.R., and Miller, H.D. *The Theory of Stochastic Processes.* New York: Wiley, 1968.

Davenport, W.B., Jr. *Probability and Random Processes.* New York: McGraw-Hill, 1970.

Denning, P.J., and Buzen, J.P. "The Operational Analysis of Queueing Network Models." *Computing Surveys* 10 (1978): 225–61.

Drake, A.W. *Fundamentals of Applied Probability.* New York: McGraw-Hill, 1967.

Erlander, S. "The Remaining Busy Period for a Single Server Queue with Poisson Input." *Operations Research* 13 (1965): 734–46.

Feller, W. *An Introduction to Probability Theory and Its Applications, 3rd ed.* New York: Wiley, 1968.

——. *Probability Theory and Its Applications, volume 2.* New York: Wiley, 1966.

Fishman, G.S. "Estimation in Multiserver Queueing Simulations." *Operations Research* 22 (1974): 72–78.

Fishman, G.S., and Kao, E.P.C. "A Procedure for Generating Time-dependent Arrivals for Queueing Simulation." *Naval Research Logistics Quarterly* 24 (1977): 661–66.

Gordon, G. *System Simulation.* Englewood Cliffs, N.J.: Prentice-Hall, 1969.

Gross, D., and Harris, C.M. *Fundamentals of Queueing Theory.* New York: Wiley, 1974.

Gue, R.L., and Thomas, M.E. *Mathematical Methods in Operations Research.* New York: Macmillan, 1968.

Hillier, F.S., and Lieberman, G.J. *Operations Research, 2nd ed.* San Francisco: Holden-Day, 1974.

Hoel, P.G.; Port, S.C.; and Stone, C.J. *Introduction to Stochastic Processes.* Boston: Houghton-Mifflin, 1972.

IBM Data Processing Techniques Manual (F20-0007-1). *Analysis of Some Queueing Models in Real-time.* Second edition (September 1971).

Karlin, S., and McGregor, J. "Many Server Queueing Processes with Poisson Input and Exponential Service Times." *Pacific Journal of Mathematics* 8 (1958): 87–118.

Karlin, S. *A First Course in Stochastic Processes.* New York: Academic Press, 1968.

Kendall, D.G. "Some Problems in the Theory of Queues." *Journal of the Royal Statistical Society, Series B* 13 (1951): 151–85.

——. "Stochastic Processes Occurring in the Theory of Queues and Their Analysis by Means of the Imbedded Markov Chain." *Annals of Mathematical Statistics* 24 (1953): 338–54.

Kleinrock, L. *Queueing Systems. Volume I: Theory.* New York: Wiley, 1975.

——. *Queueing Systems. Volume II: Computer Applications.* New York: Wiley, 1975.

Koenigsberg, E. "Finite Queues and Cyclic Queues." *Operations Research* 8 (1960): 246-53.

——. "Queueing with Special Service." *Operations Research* 4 (1956): 212-20.

Kotiah, T.C.T.; Thompson, J.W.; and Waugh, W.A.O'N. "Use of Erlangian Distributions for Single-server Queueing Systems." *Journal of Applied Probability* 6 (1969): 584-93.

Lindley, D.V. "The Theory of Queues with a Single Server." *Proceedings of the Cambridge Philosophical Society* 48 (1952): 277-89.

Little, J.D.C. "A Proof of the Queueing Formula L = λW." *Operations Research* 9 (1961): 383-87.

Maisel, H., and Gnugnoli, G. *Simulation of Discrete Stochastic Systems.* Chicago: Science Research Associates, 1972.

Martin, J. *Design of Real-time Computer Systems.* Englewood Cliffs, N.J.: Prentice-Hall, 1967.

Miller, R.G., Jr. "Priority Queues." *Annals of Mathematical Statistics* 31 (1960): 86-101.

Morse, P.M. *Queues, Inventories, and Maintenance.* New York: Wiley, 1958.

Morse, P.M.; Garbor, H.N.; and Ernst, M.L. "A Family of Queueing Problems." *Operations Research* 2 (1954): 444-45.

Naylor, T.H.; Balintfy, J.L.; Burdick, D.S.; and Chu, K. *Computer Simulation Techniques.* New York: Wiley, 1966.

Panico, J.A. *Queuing Theory: A Study of Waiting Lines for Business, Economics, and Science.* Englewood Cliffs, N.J.: Prentice-Hall, 1969.

Papoulis, A. *Probability, Random Variables, and Stochastic Processes.* New York: McGraw-Hill, 1965.

Parzen, E. *Modern Probability Theory and Its Applications.* New York: Wiley, 1960.

——. *Stochastic Processes.* San Francisco: Holden-Day, 1962.

Phillips, D.T. "Applied Goodness of Fit Testing." *Operations Research, Momograph Series,* No. 1, AIIE-OR-72-1. Atlanta: American Institute of Industrial Engineers, 1972.

Prabhu, N.U. *Queues and Inventories: A Study of Their Basic Stochastic Processes.* New York: Wiley, 1965.

Rider, K.L. "A Simple Approximation to the Average Queue Size in the Time-dependent M/M/1 Queue." *Journal of the ACM* 23 (1976): 361-67.

Riordan, J. *Stochastic Service Systems.* New York: Wiley, 1962.

Ross, S.M. *Introduction to Probability Models.* New York: Academic Press, 1972.

Saaty, T.L. *Elements of Queueing Theory with Applications.* New York: McGraw-Hill, 1961.

——. "Resume of Useful Formulas in Queueing Theory." *Operations Research* 5, (April 1957): 162-200.

Sasieni, M.; Yaspan, A.; and Freedman, L. *Operations Research: Methods and Applications.* New York: Wiley, 1959.

Sauer, C.H., and Chandy, K.M. "Approximate Solution of Queueing Models." *Computer* 13 (1980): 25-32.

Shanthikumar, J.G. "On a Single-server Queue with State-dependent Service." *Naval Research Logistics Quarterly* 26 (1979): 305–09.

Smith, W.L. "On the Distribution of Queueing Times." *Proceedings of the Cambridge Philosophical Society* 49 (1957): 449–61.

Stidman, S., Jr. "A Last Word on $L = \lambda W$." *Operations Research* 22 (1974): 417–21.

Takacs, L. *Introduction to the Theory of Queues.* New York: Oxford University Press, 1962.

——. "Priority Queues." *Operations Research* 12 (1964): 63–74.

——. "A Single-server Queue with Poisson Input." *Operations Research* 10 (1962): 338–97.

——. "The Time Dependence of a Single-server Queue with Poisson Input and General Service Times." *Annals of Mathematical Statistics* 33 (1962): 1340–48.

——. "Transient Behavior of Single-server Queueing Process with Recurrent Input and Exponentially Distributed Service Times." *Operations Research* 8 (1960): 231–45.

Wagner, H.M. *Principles of Operations Research, 2nd ed.* Englewood Cliffs, N.J.: Prentice-Hall, 1975.

Wiederhold, G. *Data Base Design.* New York: McGraw-Hill, 1977.

Zussman, R. "Computer Queueing Analysis on a Handheld Calculator." *Computer Design* (November 1977): 85–94.

——. "Predicting Queue Performance on a Programmable Handheld Calculator." *Computer Design* (August 1978): 93–100.

Index

Absolute frequency, 9-10, 12, 18
Allen, A.O., 90, 96,110, 126, 134
Average
  customer arrival rate. See λ
  number of customers
    in queue. See L_q
    in system. See L
  service rate. See μ
  time
    for service. See W
    in queue. See W_q

BASIC, 5, 26, 28, 46
Balance equation, 35
Bernoulli trial, 48-49
Binomial theorem, 34-37
Birth-death process, 34-37

c (number of servers), 73, 95
  Kendall notation, 91-92
  M/M/c, 110-15, 117-18, 120-22
  multiserver queueing system, 89
Central limit theorem, 31
Chang, W., 34, 81, 126
Clarke, A.B., 34, 59, 134

Continuous. See Mean; Probability
  distribution function; Random
  variables; Variance
Cox, D.R., 79
Cumulative distribution function, 7-10,
  18-22, 27, 40, 43
  exponential, 63-64
  normal, 30
  uniform, 41, 45-46
Customer, 3, 74, 84, 91, 93, 112
  arrival patterns, 75-79, 89, 96
  behavior, 81-82
C_x^2 (squared coefficient of variation),
  90

Discrete. See Mean; Probability density
  function; Random variables;
  Variance
D/D/1. See $P_0(t)$; $P_n(t)$; Queueing
  systems
Erlang. See Mean; Probability density
  function; Probability distribution
  function; Variance
Erlang, A.K., 4, 67, 123-24
Event, 7, 11-12, 53-57
Expected value. See Mean